Low-Cost Urban Sanitation

Duncan Mara
Department of Civil Engineering, University of Leeds, U.K.

JOHN WILEY & SONS
Chichester · New York · Brisbane · Toronto · Singapore

National 01243 779777
International (+44) 1243 779777

Other Wiley Editorial Offices

John Wiley & Sons, Inc., 605 Third Avenue,
New York, NY 10158-0012, USA

Jacaranda Wiley Ltd, 33 Park Road, Milton,
Queensland 4064, Australia

John Wiley & Sons (Canada) Ltd, 22 Worcester Road,
Rexdale, Ontario M9W 1L1, Canada

John Wiley & Sons (Asia) Pte Ltd, 2 Clementi Loop #02-01,
Jin Xing Distripark, Singapore 0512

Library of Congress Cataloging-in-Publication Data

Mara, D. Duncan (David Duncan), 1944 –
Low-cost urban sanitation / D. Mara.
p. cm.
Includes bibliographical references and index.
ISBN 0 471 96163 9
1. Sewerage—Tropics. 2. Sewerage—Economic aspects—Tropics.
3. Sewerage—Social aspects—Tropics. 4. Sanitation—Tropics.
I. Title.
TD626.5.M37 1996
628′.2′091732—dc20 95-42464
CIP

British Library Cataloguing in Publication Data

A catalogue record for this book is available from the British Library

ISBN 0 471 96163 9

For

John Kalbermatten

John Maurice Kalbermatten, whose career at the World Bank spanned 1971–1986, was the first to realise that investments in sewerage were not reaching the urban poor. He directed the World Bank Research Project on Appropriate Technologies for Water Supply and Sanitation in Developing Countries (1976–1978), which was the essential multidisciplinary precursor to the International Drinking Water Supply and Sanitation Decade (1981–1990). He established, with funds from the United Nations Development Programme, the World Bank – UNDP Sanitation Technology Advisory Group (TAG), which did so much during 1978–1986 to develop and promote affordable and sustainable urban sanitation in developing countries. He retired from the World Bank in June 1986 as its Senior Water and Wastes Adviser.

To work with John Kalbermatten, as I did during 1978–1986, was a challenge, an inspiration and a privilege. If anyone can be described as the Edwin Chadwick of our times, John Kalbermatten most deservedly can.

Contents

Preface

Tropical public health engineering is an exciting field. To meet the challenges of the International Drinking Water Supply and Sanitation Decade (1981–1990), and now Safe Water 2000, which is the international successor to the IDWSSD (and which, despite its title, does encompass sanitation), developing countries, together with international agencies and donor countries, must invest in their water supply and sanitation sector, not only directly but also through programmes of human resource development. This book is a contribution to that development. It is primarily addressed to undergraduate and postgraduate students of civil engineering and tropical public health engineering, but it is hoped that practising engineers will also find it of use.

Low-cost urban sanitation is not just a question of technologies. These are obviously important, but so too is an understanding of excreta-related disease transmission, and how this transmission can be controlled by sanitation interventions. The human dimension cannot be ignored: sanitation systems must be socioculturally acceptable; people must know how to use and maintain them, and why this is important. Sanitation options have also to be economically appropriate and financially affordable, and government agencies and non-governmental organizations (NGOs) responsible for, and working in, the sector need to work imaginatively with low-income urban communities so that they get the sanitation facilities that they so desperately need. If this book helps to stimulate these activities, it will have been a worthwhile endeavour.

Duncan Mara
Leeds, August 1995

Acknowledgements

Many people have helped me over the years in my work on low-cost sanitation, and to them I owe a considerable debt of gratitude: John Kalbermatten, to whom this book is dedicated, Dr DeAnne Julius and Richard Middleton, all formerly at the World Bank; Professor Richard Feachem, Professor David Bradley, Dr Sandy Cairncross, Dr Astier Almedom and Dr Peter Kolsky, of the London School of Hygiene and Tropical Medicine; Dr Michael McGarry, of Cowater International, Ottawa; Dr Peter Morgan, of the Mvuramanzi Trust, Harare; Piers Cross, of the Mvula Trust, Johannesburg; Dra Maria Helena Marecos do Monte, of the Laboratório Nacional de Engenharia Civil, Lisbon; Eng. Augusto Sergio Guimarães, of GAIA Engenharia Ambiental, Rio de Janeiro; Dr Richard Otis, of Owen Ayres & Associates, Madison WI; Dr Howard Pearson, University of Liverpool; and Professor Gerrit Marais, formerly of the University of Cape Town.

Docendo dedici: several former masters and doctoral students have broadened my intellectual sanitation knowledge base: Dr Gehan Sinnatamby, Dr Graham Alabaster and Dr Steve Greenhalgh, of the United Nations Centre for Human Settlements, Nairobi; Professor Cicero Onofre Neto, of the Universidade Federal do Rio Grande do Norte, Natal; Dr Rachel Ayres, Dr Rebecca Stott and Susanne Niedrum, of the University of Leeds; and Anthony and Juliet Waterkeyn, of WaterAid, Harare.

The sanitation technology algorithm given in Figure 13.1 is based in part on the one we developed in *Appropriate Sanitation Alternatives: A Planning and Design Manual* (Johns Hopkins University Press, 1982), which was modified by Jeff Broome when he was working here in Leeds on the 1986 World Bank tape–slide programme, *Sanitation Technology Selection*. This, in turn, was recently modified by Jacqui Hamer as part of her final year civil engineering project work in Leeds in 1995.

There are two other persons whom I would like to thank (and indeed whom I should have thanked many times before): firstly, Dr Tony Cusens, formerly Professor of Civil Engineering at the Universities of St. Andrews, Dundee and Leeds, who provided over a period of some 25 years an academic environment that not only permitted but actively encouraged my overseas work on urban sanitation; and secondly, Dr J. Kevin Newman, Professor of Classics at the University of Illinois at Urbana-Champaign, who taught me not only Latin and Greek but also, and more importantly, how to think. He first introduced me to the dictum of Callimachus of Cyrene (c. 305–c. 240 BC), which has shaped my writing ever since:

μέγα βιβλίον μέγα κακόν

Copyright Acknowledgements

I am most grateful to several colleagues and institutions for permission to include copyright material. These are: the American Society of Civil Engineers (Figures 6.4 and 6.7); Dr D. Barnes, University of New South Wales (Figure 9.5); Brain Industries Ltd (Figures 7.1, 7.2 and 7.3); Professor D. W. T. Crompton, University of Glasgow (Figure 2.5); Her Majesty's Stationery Office (Figures 3.3 and 3.7); Ifö Sanitar AB, Bromolla (Figure 4.7); Jane Starritt (Figures 11.1 and 11.2); Sulabh International Social Service Organization (Figure 1.4); Thomas Telford Ltd (Figure 1.1); Waste Consultants, Gouda (Figure 7.4); WaterAid (Figure 1.2); the World Bank (Figures 2.1, 2.2, 2.4, 3.1, 3.6, 3.7, 3.10, 3.14, 3.16, 4.1, 4.3, 4.4, 4.9, 4.10, 8.1, 8.3, 8.4, 8.5, 15.1 and Table 15.2); and the World Health Organization (Figures 2.3 and 5.3).

1

Urban Sanitation Needs

1.1 GLOBAL NEEDS IN URBAN SANITATION

There was considerable progress in the provision of low-cost water supply and sanitation to low-income communities during the International Drinking Water Supply and Sanitation Decade (1981–1990), but much still remains to be done in the 1990s and doubtless beyond (Figure 1.1). Population growth is likely to continue to equal or exceed any increase in the numbers of people served, such that the numbers of those requiring improved water supplies and sanitation will continue to increase, almost relentlessly so. (The world's urban population was 2.3 billions in 1990, and the United Nations estimates

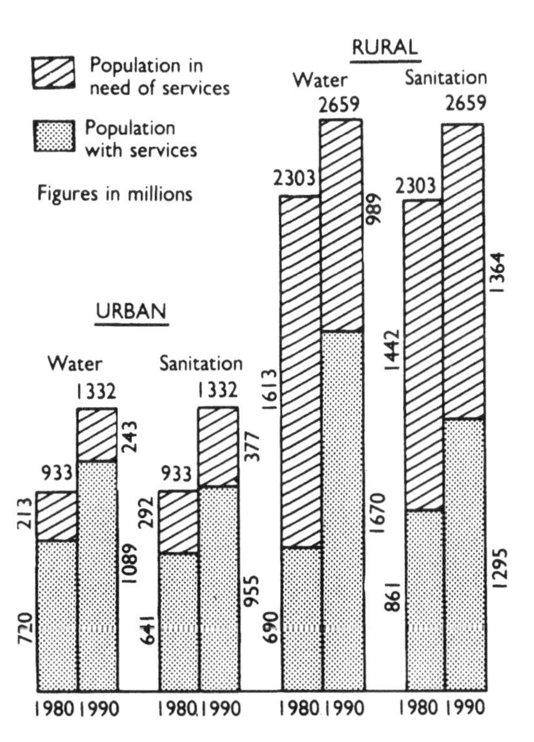

Figure 1.1 The success of the International Drinking Water Supply and Sanitation Decade (1981–1990) and the challenge for Safe Water 2000

that by 2020 it will have doubled to 4.6 billions, with 93 percent of this growth occurring in developing countries.)

The health burden borne by those without adequate water supplies and sanitation is huge: for example, the World Health Organization estimates that some 12 000 000 people die each year from water- and excreta-related diseases, including 4 000 000 children under the age of five who die from diarrhoeal disease alone. More startling than these mortality figures is the WHO's estimate that fully 80 percent of all morbidity in developing countries is due to water- and excreta-related diseases. These figures are for both urban and rural areas, but those most at risk are the urban poor, especially children (Figure 1.2).

The World Bank has estimated that water- and excreta-related diseases were responsible in 1979 for the loss of around 400 billion working days in Africa, Asia and Latin America. At US$ 0.50 per day, this loss amounted to some US$ 200 billion. Since its gross national product (GNP) was then about US$ 370 billion, the output of the developing world was below its productive potential by as much as 200/(370 + 200), i.e. 35 percent. So investments in improved water supplies and improved sanitation can bring benefits not only to individuals, but also to national economies. Governments need, therefore, to invest in their water supply and sanitation sector. If they do not, their economies will not

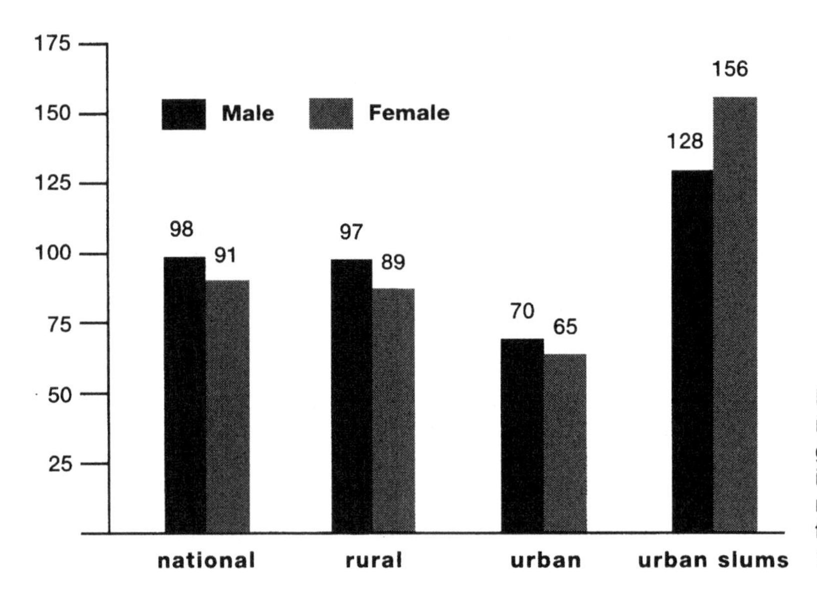

Figure 1.2 Infant mortality rates in Bangladesh in 1991. (IMR is the number of children who die in their first year of life, per 1000 live births)

develop as productively as possible, and people will continue—in the words of the late Barbara Ward—to defecate themselves to death.

1.2 TECHNOLOGICAL OPTIONS FOR SANITATION

Conventional sewerage, for so long considered by engineers and planners (also by politicians) as the *only* sanitation technology for developing country cities, has two major disadvantages:

- extreme high cost: the World Bank found sewerage investment costs in eight capital cities in developing countries to be US$ 600–4000 per household (1980 prices), with total annual costs per household (see Chapter 12) of US$ 150–650, and
- the need for in-house (or at least on-plot) water.

It is thus simply not an option for low-income urban communities, unless they are massively subsidized. Fortunately, several well-tried and robust alternative sanitation technologies exist that are not only cheaper than conventional sewerage, but are also able to deliver the same health benefits and offer the opportunity for community participation (self-help) to reduce costs. These are:

On-site systems: VIP latrines (Chapter 3)
Pour-flush toilets (Chapter 4)
Septic tanks (Chapter 6)

Off-site systems: Settled sewerage (Chapter 8)
Simplified sewerage (Chapter 9)

With VIP latrines and pour-flush toilets, arrangements have to be made for sullage disposal (Chapter 5)—sullage is the non-toilet wastewater generated by a household (i.e. the waste from sinks and showers). In high-density urban areas simplified sewerage can be cheaper than on-site systems (see Figure 9.2), but off-site systems have the requirement for effective sewage treatment—"effective treatment" in this context means treatment to kill or remove excreted pathogens (Chapters 2 and 10). Indeed, one can argue that VIP latrines and pour-flush toilets whose pits are dry (i.e. do not penetrate the groundwater table) have better health benefits than sewerage, even unconventional sewerage: pathogen travel is minimal from dry pits; so once deposited there, excreted pathogens stay there and present no risk to health. In contrast, pathogen travel with sewer systems is vast and, if treatment is not effective (or, as all too often, absent), the resulting widespread environmental contamination with excreted pathogens can pose significant health risks.

1.2.1 Communal or individual sanitation?

Communal sanitation systems generally suffer from the lack of maintenance and are often, as a result, in a high state of faecal disorder and odour (Figure 1.3). There are, of course, the occasional exceptions to this: for example, the excellent communal sanitation blocks in some Indian cities managed by Sulabh International Social Service Organization (Figure 1.4):

Figure 1.3 A poorly-maintained communal toilet in a West African city. Note the faeces on the ground near the toilet: faecal density and distance from the toilet almost seem to be related by an inverse square law!

Figure 1.4 Communal sanitation block in peri-urban India managed by the non-governmental organization, Sulabh International Social Service Organization

women and children are allowed free entry, but men pay 50–100 paise (1.5–3 US cents). There are hand-washing facilities and showers (soap is also provided), as well as pour-flush toilets.

Individual sanitation facilities (one per household or per compound) are, however, much better, and should be the norm in urban sanitation programmes. The health benefits of sanitation (see Chapter 2) are much better able to be realized with individual, rather than communal, sanitation (see also Chapter 14). But individual sanitation facilities need good operation and maintenance at the household level: one might term this HLOM—household-level operation and maintenance—and household members (especially women) may need training in HLOM techniques (Chapter 14).

If it really is impossible to provide individual sanitation (for example, in palafitic areas—houses built on stilts over water), then one solution would be to have a communal sanitation block in which each household has its own toilet cubicle (and the key to it) and is responsible for its maintenance. This could even be done on a per compound (rather than a per household) basis, but this can lead to problems, especially if several unrelated families live in the same compound.

1.3 COMPLEMENTARY INPUTS

Sanitation by itself is not enough to improve health; in mathematical terminology, it is necessary but not sufficient. There are other requirements: for example, improvements in water supply (especially, but not only, to combat the waterwashed transmission of Category I and II excreta-related diseases—see section 2.2) and hygiene education (Chapter 14).

Prior to implementation of the chosen sanitation technology (see Chapter 13), proper planning must take place (Chapter 15) and this involves comprehensive sociocultural and socioeconomic evaluations (Chapters 11 and 12).

Costs have to be recovered, and mechanisms for cost recovery must be equitable and not place a significant financial burden on low-income groups (Chapter 12). If necessary, there should be cross-subsidies from middle- and upper-income groups to the urban poor. What should *not* happen is that the poor should subsidize the rich. But this does happen: often a water and sewerage authority faces financial difficulty because the maximum 100 percent surcharge that it is allowed to charge for sewerage on the monthly household water bill is insufficient to cover the cost of providing the sewerage service to its middle- and upper-income customers; so it has to increase the cost of water and the poor end up paying more for their water in order that the rich can continue to be served by high-cost sewers. (Remember Mara's dictum for development: *the poor should not pay for what the rich get free*.)

1.4 INTEGRATED URBAN DEVELOPMENT

Often sanitation improvements are better done as part of a programme of integrated urban development, as so many sectors contributing to urban health and well-being require improvement. These include legal aspects (for example, land title); environmental services, such as water supply, drainage and garbage collection; improvements to housing, access roads and transportation; provision of primary health care facilities, crèches and schools; and employment creation. Community involvement in urban development planning generally means that community acceptance of infrastructure improvements, including sanitation, is much higher. This is especially true in low-income areas, and the municipal authorities must work with the urban poor to improve their overall quality of life, so that they can more effectively contribute to urban socio-economic development.

1.5 FURTHER READING AND INFORMATION

S. Cairncross, *Sanitation and Water Supply: Practical Lessons from the Decade*. Water and Sanitation Discussion Paper No. 9. The World Bank, Washington, DC (1992).

A. Cotton and R. Franceys, *Services for Shelter*. Liverpool University Press, Liverpool (1991).

D. Drakakis-Smith, *The Third World City*. Methuen, London (1987).

J. E. Hardoy, D. Mitlin and D. Satterthwaite, *Environmental Problems in Third World Cities*. Earthscan Publications, London (1992).

W. Hogrewe, S. D. Joyce and E. A. Perez, *The Unique Challenges of Improving Peri-urban Sanitation*. Report No. 86. Water and Sanitation for Health Project, Arlington, VA (1993).

J. M. Kalbermatten, D. S. Julius and C. G. Gunnerson, *Appropriate Sanitation Alternatives: A Technical and Economic Appraisal*. Johns Hopkins University Press, Baltimore, MD (1982).

J. M. Kalbermatten, D. S. Julius, C. G. Gunnerson and D. D. Mara, *Appropriate Sanitation Alternatives: A Planning and Design Manual*. Johns Hopkins University Press, Baltimore, MD (1983).

C. Kessides, *The Contributions of Infrastructure to Economic Development*. World Bank Discussion Paper No. 213. The World Bank, Washington, DC (1993).

J. A. Lee, *The Environment, Public Health and Human Ecology*. Johns Hopkins University Press, Baltimore, MD (1985).

Lessons Learned in Water, Sanitation and Health: Thirteen Years of Experience in Developing Countries, updated edition. Water and Sanitation for Health Project, Arlington, VA (1993).

D. W. Pearce and J. J. Warford, *World Without End: Economics, Environment and Sustainable Development*. Oxford University Press, New York (1993).

J. Pickford, *Low-Cost Sanitation: A Survey of Practical Experience*. Intermediate Technology Publications, London (1995).

Prospects of World Urbanization 1988. United Nations Department of International Economic and Social Affairs, New York (1988).

United Nations Development Programme, *Safe Water 2000: The New Delhi Statement*. UNDP, New York (1990).

United Nations General Assembly, *Achievements of the International Drinking Water Supply and Sanitation Decade*. Report of the Secretary General A/45/327 (13 July). United Nations, New York (1990).

The World Bank has produced three videotapes and fifty tape-slide programmes on water supply and sanitation in developing countries, under the general title *Information and Training for Low-cost Water Supply and Sanitation*. It also produces a series of *Water and Sanitation Reports* and *Discussion Papers*. Further details are available from:

Water and Sanitation Division, TWU
The World Bank
1818 H St NW
Washington, DC 20433
USA

Fax: +1 202 522 3228

IT Publications produce the quarterly journal *Waterlines*, which covers the field of low-cost water supplies and sanitation in developing countries. IT also issues an extremely useful annual book catalogue. Further details are available from:

IT Publications Ltd
103–105 Southampton Row
London WC1B 4HH
England

Fax: +44 171 436 2013

The following organizations also produce catalogues or listings of relevant publications:

World Health Organization
1211 Geneva 27
Switzerland

Fax: +41 22 791 0746

UNICEF
3 UN Plaza
New York, NY 10017
USA

Fax: +1 212 779 1679

IRC International Water and Sanitation Centre
PO Box 93190
2509 AD, The Hague
The Netherlands

Fax: +31 70 381 4034

Water and Sanitation in Developing Countries (SANDEC)
(formerly the International Reference Centre for Wastes Disposal)
Ueberlandstrasse 133
Duebendorf
Switzerland

Fax: +41 1 823 5399

Environmental Health Project
1611 N. Kent Street, Suite 300
Arlington, VA 22209-2111
USA

Fax: +1 703 253 9004

(EHP is the successor to WASH, Water and Sanitation for Health. WASH reports are available from EHP.)

GARNET—the Global Applied Research Network in Water Supply and Sanitation—coordinates research on low-cost sanitation and issues a *Newsletter* (which is also available on e-mail). Further details are available from:

The GARNET Coordinator
WEDC
Loughborough University of Technology
Leicestershire LE11 3TU
England

Fax: +44 1509 211079
e-mail: WEDC@lut.ac.uk

2

Sanitation and Health

2.1 THE IMPROVEMENT OF PUBLIC HEALTH

The principal purpose of programmes to improve urban sanitation is to improve public health. Usually, such programmes are mainly concerned with sanitation improvements in low-income communities, and the health profile of these communities is very poor (Figure 1.2). If we are to improve public health through improved sanitation, then we need to have a clear understanding of the diseases that are most prevalent when sanitation is poor, and how they are transmitted. These diseases are the excreta-related diseases, and they are caused by viruses, bacteria, protozoa and helminths (worms). An alphabetical list of the 50–60 excreta-related diseases is not especially helpful, nor is one that groups them into viral, bacterial, protozoan and helminthic diseases. They need to be divided into groups of diseases that have common environmental transmission routes. Such a classification is called an *environmental classification*, and the environmental classification of excreta-related diseases presented below was developed by Professor Richard Feachem and his colleagues at the London School of Hygiene and Tropical Medicine in the early 1980s.

2.2 THE ENVIRONMENTAL CLASSIFICATION OF EXCRETA-RELATED DISEASES

There are two different ways in which excreta can cause infection (Figure 2.1). In the first, the pathogens in the excreta of one person reach another person in whom they initiate infection. These are the *excreted infections*, and they are described in subsection 2.2.1. They comprise the first five categories of excreta-related diseases in the environmental classification developed in subsection 2.2.2.

The second way in which excreta can cause infection is that they may promote the growth of insects such as flies, mosquitoes and cockroaches, which act as vectors of excreted and other diseases; or rodents such as rats may be involved in the transmission of excreted and other diseases. These insect-vector and rodent-vector diseases form the last two categories of excreta-related diseases, and they are also described in subsection 2.2.2.

2.2.1 Excreted infections

For the successful transmission of an excreted infection, the pathogen causing that infection has to pass from the excreta of one person to the mouth or other port of entry of another person. Successful transmission depends on how many pathogens are excreted by the first person, how these numbers change during transmission in the environment, and how probable it is that the second person becomes infected.

The number of pathogens excreted is termed the *excreted load*. How this number changes during its environmental transmission is governed by three key properties of the pathogen: *latency* or how long it takes for the pathogen, once excreted, to become infective; *persistence*, or how long the pathogen can survive in the environment; and *multiplication*, the ability of the pathogen to increase its numbers whilst in the environment. A fourth key property of the pathogen is its *infectivity*, which is the probability of one organism initiating infection. These five concepts are discussed below, and their relation is shown in Figure 2.2.

Excreted load

The number of pathogens excreted by an infected person varies widely. For example, a person with cholera may excrete

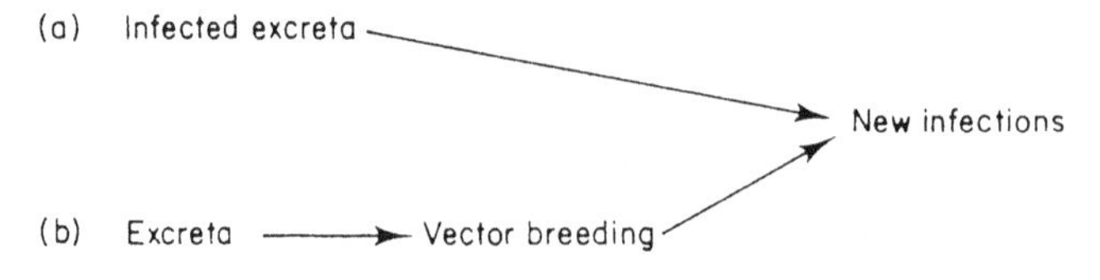

Figure 2.1 The links between excreta and infection. In (a) the excreta contain the pathogens that are transmitted by various environmental routes to a new host. In (b) the excreta, or sewage, facilitate the breeding of certain flies and mosquitoes, or encourage rodents, that can act as vectors of excreted and other pathogens

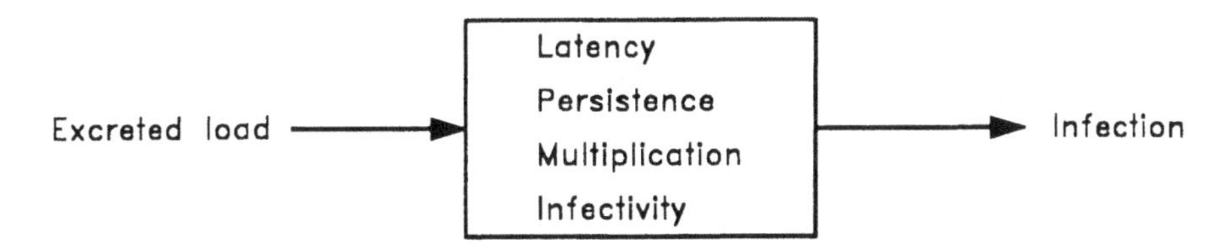

Figure 2.2 The principal factors that affect the successful transmission of an excreted pathogen

some 10^{13} cholera vibrios a day. Someone with a light infection of *Ascaris lumbricoides*, the human roundworm, may excrete a few hundred thousand eggs per day (each female worm can produce up to 200 000 eggs per day).

The excreted load depends on the state of infection: as the cholera victim, for example, becomes better, the number of vibrios excreted falls, eventually, of course, to zero. Another good example is schistosomiasis (commonly also called bilharzia): infected children, who generally show few clinical signs of the disease, excrete large numbers of schistosome eggs, whereas adults in the terminal stage of the disease generally produce few or no eggs.

Latency

Latency is the interval between the excretion of a pathogen and its becoming infective to a second person. Many excreted pathogens—including all viruses, bacteria and protozoa—are non-latent: that is, they are infective immediately they are excreted. Latency is an important property, therefore, only of the helminths, and only three helminths are non-latent (these are the human pinworm or threadworm, *Enterobius vermicularis*; the dwarf tapeworm, *Hymenolepis nana*; and the small nematode, *Strongyloides stercoralis*, which can be both non-latent and latent). All other excreted helminths are latent, and their latency varies from a few days to a few weeks. During their latent period the worm changes from a non-infective form to its infective form. This development may occur wholly in the environment outside the body (as in the case of the geohelminths, which are responsible for the third category of excreta-related diseases), or it may take place partly in the environment and partly in an intermediate host (a cow or pig in the case of the tapeworms; or a water snail and possibly also a fish or an aquatic vegetable in the case of the water-based trematode worms; these are the fourth and fifth categories of excreta-related diseases, respectively).

Persistence

How long an excreted pathogen can survive in the environment outside the body is the property most indicative of the health hazard it poses. A pathogen that is very persistent—for example, *Ascaris* eggs, which can survive for many months, even years—will be a risk in sewage and sewage treatment processes and during reuse of the treated effluent for crop irrigation (see Chapter 10). Even excreted bacteria, which survive generally only for a few weeks, can also constitute a risk in this way.

Excreted pathogens that have only a short or very short persistence have to reach their new host very quickly. Their transmission cannot be a long one of the type just described for *Ascaris* eggs and excreted bacteria: it has to be short and commonly occurs within the immediate household environment, from one family member to another, or to a neighbour or visitor, generally through poor personal and domestic hygiene.

Multiplication

Some excreted pathogens can, given the right environmental conditions, multiply—often several thousandfold or several millionfold: for example, excreted bacteria in food and milk, and the water-based trematodes in aquatic snails. Thus, a low excreted load can readily multiply to increase the risk of infection. Excreted viruses and excreted protozoa cannot multiply, and therefore for them to be able to be transmitted successfully, their infectivity has to be very high.

Infectivity

Knowledge about infectivity—the probability of infection from one organism—is far from perfect. What information there is has usually come from volunteer studies: a pathogen is given to groups of volunteers who are then monitored to see if they become infected (and, if they do, they are then quickly treated). Generally, the volunteers have been healthy adults from non-endemic areas, and their response is very different from that of malnourished children in the urban tropics as is discussed below. Nevertheless, we cannot ignore infectivity, however imperfect our knowledge. In general terms, we can use the following descriptive categories for the probability of infection from one organism:

High infectivity:	$<10^{-2}$
Medium infectivity:	10^{-2}–10^{-6}
Low infectivity:	$>10^{-6}$

Host susceptibility

The excreta of one person will cause disease in another person once infected, but only if that person is susceptible. Host susceptibility governs the severity of the disease: a person may be susceptible or, due to previous exposure or immunization, immune or have a varying degree of resistance. Small children are often most at risk from excreted infections: diarrhoeal diseases, for example, occur mainly in one to four year olds, with older children and adults at progressively lower risk (Figure 2.3). Thus, urban sanitation programmes must recognize not only that children are at most risk, but also that they are the main source of many excreted infections: hygiene education is extremely important here as the commonly-held

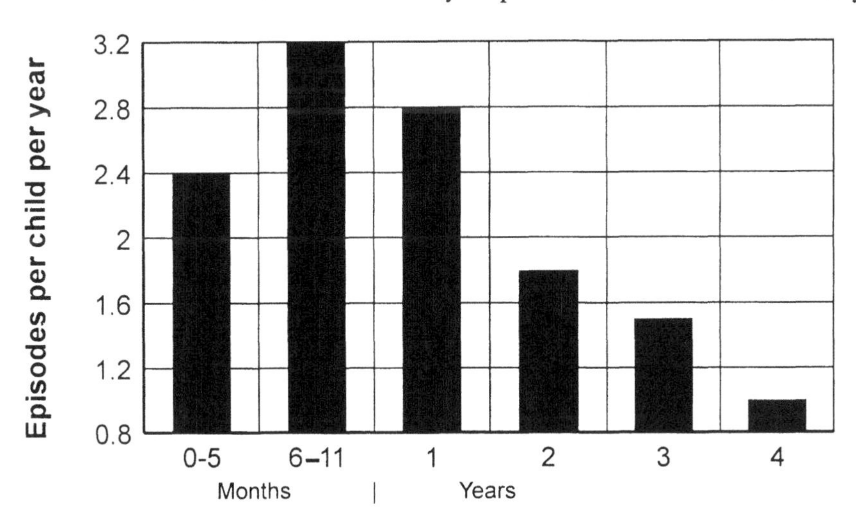

Figure 2.3 Estimated median diarrhoeal disease morbidity rates for children under five in developing countries, by age group (based on 15 studies in developing countries). Note that those most at risk are children in the first two years of life

belief that children's excreta are harmless is simply not true—mothers need to be taught that they are extremely hazardous: the excreta of children too young to use a toilet should be disposed of safely; and once children are old enough to use a toilet, they should be taught to do so (see Chapter 14). However, it must also be recognized that not all hosts are human.

Non-human hosts

While some excreted infections are infections only of humans—for example, shigellosis, or bacillary dysentery, caused by bacteria of the genus *Shigella*—others involve animals. This involvement is of two types (Figure 2.4). In the first, wild or domestic vertebrates act as an alternative reservoir of infection—examples are salmonellosis, caused by bacteria of the genus *Salmonella*, which is also an infection of poultry, cattle, pigs and rodents (as well as of several domestic pets; for example, cats and dogs); and campylobacteriosis, caused by bacteria of the genus *Campylobacter*, which is also an infection of cattle and poultry (and other animals, also puppies and kittens). Transmission may be animal-to-animal or animal-to-person or person-to-person. This type of disease is called a zoonosis. It is not easy to control zoonoses by sanitation, since sanitation only controls person-to-person transmission; animal-to-person transmission may still occur and this can be controlled only by other means—for example, pet and chicken hygiene or rodent control.

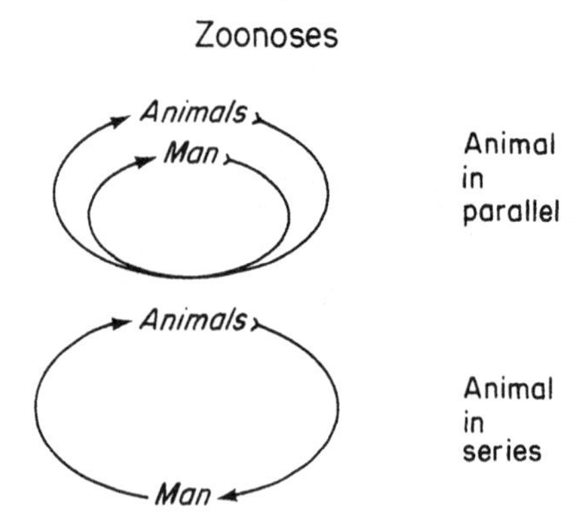

Figure 2.4 Ways in which animals are involved in the transmission of human excreted infection. An example of 'animals in parallel' is poultry (especially chickens), which transmit the bacterial diseases salmonellosis and campylobacteriosis. Examples of 'animals in series' are cows and pigs, the intermediate hosts of beef and pork tapeworm; and snails, the intermediate aquatic host of the water-based helminthiases

Animals may be involved in a second way, as intermediate hosts. Examples are cows and pigs in the case of beef and pork tapeworm disease; aquatic snails in the case of schistosomiasis, and aquatic snails and fish (usually carp or crayfish) in the case of oriental liver fluke infection or clonorchiasis. Sanitation can be very effective in controlling these diseases since it can prevent human excreta reaching the animal hosts; but it may not always be effective in so doing: if sewage is not properly treated (see Chapter 10) before being used to irrigate pasture, for example, cows can become infected with beef tapeworm eggs; or improperly treated sewage can be discharged into water and schistosome eggs can hatch and infect snails.

2.2.2 Categories of excreta-related diseases

The environmental classification of excreta-related diseases is presented in Table 2.1. It contains seven categories: categories I–V are the excreted infections, and categories VI and VII are the insect-vector and rodent-vector excreta-related diseases, respectively.

The five categories of excreted infections are distinguished by latency, persistence and multiplication, by infectivity and by what sort of intermediate host (if any) is involved.

Category I: Non-bacterial faeco-oral diseases

The term "faeco-oral" is used to describe the beginning and end of the excreted pathogen's transmission route: it leaves one person in his or her faeces and it enters another person through his or her mouth. This category includes all the excreted viral and protozoan diseases and two helminthic diseases. The transmission features of these excreted pathogens are:

- non-latent
- low to medium persistence
- unable to multiply
- high infectivity
- no intermediate host

These infections are mainly spread in a very direct person-to-person way whenever personal and domestic hygiene is poor.

Table 2.1 Environmental classification of excreta-related diseases

Category	Environmental transmission features	Major examples of infection	Environmental transmission focus
I. Non-bacterial faeco-oral diseases	Non-latent Low to medium persistence Unable to multiply High infectivity No intermediate host	*Viral:* Hepatitis A and E Rotavirus diarrhoea *Protozoan:* Amoebiasis Crystosporidiasis Giardiasis *Helminthic:* Enterobiasis Hymenolepiasis	Personal Domestic
II. Bacterial faeco-oral diseases	Non-latent Medium to high persistence Able to multiply Medium to low infectivity No intermediate host	Campylobacteriosis Cholera Pathogenic *Escherichia coli* infection Salmonellosis Shigellosis Typhoid Yersiniosis	Personal Domestic Water Crops
III. Geohelminthiases	Latent Very persistent Unable to multiply No intermediate host Very high infectivity	Ascariasis Hookworm infection Strongyloidiasis Trichuriasis	Peri-domestic Field Crops

IV. Taeniases	Latent Persistent Able to multiply Very high infectivity Cow or pig intermediate host	Taeniasis	Peri-domestic Field Fodder crops
V. Water-based helminthiases	Latent Persistent Able to multiply High infectivity Intermediate aquatic host(s)	Schistosomiasis Clonorchiasis Fasciolopsiasis	Water Fish Aquatic species or aquatic vegetables
VI. Excreta-related insect-vector diseases		Infections in I–III transmitted mechanically by flies and cockroaches Bancroftian filariasis transmitted by *Culex quinquefasciatus*	Peri-domestic Water
VII. Excreta-related rodent-vector diseases		Infections in I–III transmitted mechanically by rodents Leptospirosis	Peri-domestic Water

Those pathogens that have the ability to survive for several days (the protozoa, for example) may also be spread by contaminated water and through the agricultural reuse of sewage.

Category II: Bacterial faeco-oral diseases

The transmission features of the bacterial excreted pathogens are:

- non-latent
- medium to high persistence
- able to multiply
- medium to low infectivity
- no intermediate host

These infections can be transmitted in the same direct person-to-person way as Category I infections, but their greater persistence means that longer environmental transmission routes are also important—for example, water and crop contamination.

The Category I—II continuum

It is apparent that there is no real clear-cut division between Categories I and II. None is latent or has an intermediate host, but the bacteria in Category II can multiply, whereas Category I excreted pathogens cannot. Persistence varies from low-medium in I to medium-high in II, and infectivity from high in I to medium-low in II. But there are exceptions: some enteroviruses, for example, have a high persistence and some shigellae may have a high infectivity; so the distinction in Table 2.1, which is based on the type of pathogen (bacteria in II, the others in I), whilst in some ways convenient, may not always be so useful. The real distinction is between Categories I and II together, comprising the faeco-oral diseases, and Categories III–V, which are all helminthic diseases.

There are a number of Category I and II infections that are still transmitted in affluent communities in Europe, North America and Australasia, for example, even though they have high standards of water supply, sanitation and hygiene. These infections include diseases due to the following pathogens:

Viruses:	enteroviruses rotaviruses
Bacteria:	salmonellae (other than *S. typhi*) *Campylobacter* *Shigella sonnei* pathogenic *Escherichia coli* *Yersinia enterocolitica*
Protozoa:	*Giardia* *Cryptosporidium*
Helminths:	*Enterobius vermicularis*

It is difficult, therefore, to see how urban sanitation programmes in the urban tropics will eliminate these excreted infections, but they can, of course, substantially reduce their incidence.

Category III: Geohelminthiases

This category contains the geohelminths or soil-transmitted nematode worms. These are:

Ascaris lumbricoides, the human roundworm

Trichuris trichiura, the human whipworm

Ancylostoma duodenale and *Necator americanus*, the human hookworms, and

Strongyloides stercoralis, the small nematode

Their transmission features are:

- latent
- very persistent
- unable to multiply
- very high infectivity
- no intermediate host

These are extremely common infections, especially *Ascaris* and the hookworms. In low-income areas of the urban tropics prevalences are often above 50 percent (that is, half the population is infected), and prevalences above 90 percent occur frequently. The number of worms per person (the "worm burden", a measure of the intensity of infection) can also be high (Figure 2.5).

Ascaris Under ideal conditions—moist, shady soil at 22–33 °C—*Ascaris* eggs take about 10–15 days to become infective, and they can survive for months (the longest recorded persistence is seven years). Larval development takes place inside the egg, and infection is initiated when a fully developed egg is ingested. The larvae hatch in the duodenum and are carried in the blood vessels to the heart

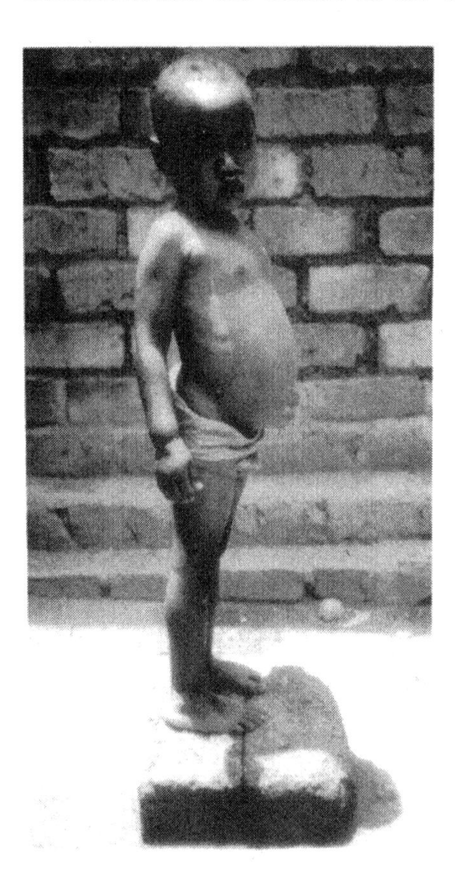

Figure 2.5 The four-year old African girl pictured above—note her distended abdomen—was given an appropriate dose of the vermifuge Levamisole. Shortly afterwards she excreted the large number of adult *Ascaris lumbricoides* worms shown

and thence to the lungs. After further development in the lungs they ascend the trachea, are swallowed and reach the small intestine where they develop into adults in two to three months. Adult worms can live for one to two years, and females lay up to 200 000 eggs a day, which leave the body in the faeces.

Hookworms In contrast to *Ascaris* eggs, hookworm eggs hatch within one to two days under suitable conditions (as for *Ascaris* eggs). After a further four to five days the larva reaches its infective form, which survives for some three to six weeks (occasionally up to 15 weeks). Infection occurs when a larva penetrates the skin of its new host, usually on the feet or ankles. Once inside the human body it travels to the heart and lungs where it partially develops, then it ascends out of the lungs, is swallowed and reaches the small intestine where it attaches itself to the intestinal wall. Adult female worms produce 5000–20 000 eggs per day, which leave the body in the faeces.

Heavy hookworm infection produces anaemia, which is particularly serious in pregnant and lactating women, and also in young children, whose mental and physical development may be retarded.

Category IV: Taeniases

This category contains the two main human cestode worms: *Taenia saginata* (the beef tapeworm) and *T. solium* (the pork tapeworm). Their transmission features are:

- latent
- persistent
- able to multiply
- very high infectivity
- cow or pig intermediate host

In the urban tropics, cows and especially pigs have ready access to human faeces, and they become infected when they eat the eggs, which are immediately infective to them. The infection is passed to humans when raw or insufficiently-well-cooked beef or pork is eaten: the larval worm attaches itself to the mucosa of the jejunum, and a mature worm develops in one to three months. Around 10^5–10^6 eggs are produced per day by each worm, and these leave the body in the faeces inside gravid segments of the worm, from which the eggs are released into the soil.

Category V: Water-based helminthiases

This category contains all the water-based human trematode worms. There are several of these, but only three are of major importance:

- *Schistosoma mansoni, S. japonicum* and *S. haematobium*, the main human schistosomes or blood flukes.
- *Clonorchis sinensis*, the oriental liver fluke (found mainly in China, Japan, Korea and Vietnam).
- *Fasciolopsis buski*, the giant human intestinal fluke (found mainly in India, Bangladesh, Thailand, Cambodia, China, Malaysia, Indonesia, Vietnam, Laos and the Philippines).

Their transmission features are:

- latent
- persistent
- able to multiply
- high infectivity
- one or two intermediate aquatic hosts: a snail (all three) and then fish (*C. sinensis*) or aquatic vegetables (*F. buski*)

The eggs are voided in the faeces (or, in the case of *S. haematobium*, in urine) and, if these reach water, they hatch to form miracidia, which enter a specific species of water snail. Huge asexual multiplication takes place inside the snail, and one to three months later aquatic larvae—cercariae for the schistosomes and metacercariae for *C. sinensis* and *F. buski*—leave the snail. The schistosome cercariae are infective to humans: infection occurs through penetration of the skin (commonly the hands, arms, legs and ankles) of anyone in contact with the contaminated water. Once inside the body, the immature flukes migrate to the lungs and thence to the portal veins where they mature. Adult females produce 10^2–10^3 eggs per day, which are voided in the faeces or, with *S. haematobium*, in the urine (many of the eggs remain in the body, where they cause complications in the liver, bladder and bowel). Urban and peri-urban schistosomiasis is common in areas with inadequate sanitation.

The metacercariae of *C. sinensis* and *F. buski* are not infective to humans. Instead, they encyst in the flesh of their second intermediate aquatic host. Human infection occurs when the cysts are consumed in raw or undercooked fish or aquatic vegetables (such as the seed pods of the water caltrop, roots of lotus, water bamboo and bulbs of water chestnuts). Once ingested, the metacercariae excyst in the duodenum. *Clonorchis* flukes mature in the bile duct, which may contain up to 6000 worms each producing 2000–4000 eggs a day. *Fasciolopsis* flukes develop in the duodenum, and each adult worm produces around 25 000 eggs per day. Animals are alternative hosts to humans for both infections: dogs for *C. sinensis*, and dogs and pigs for *F. buski*. Both diseases are important in peri-urban areas where excreta- or sewage-based aquaculture is practised (see Chapter 10).

Category VI: Excreta-related insect-vector diseases

This category contains insect-borne diseases that are excreta-related in one of two ways. Firstly, flies and cockroaches can facilitate the transmission of Category I–III infections by carrying viruses, bacteria and nematode eggs on their bodies, or in their intestinal tract, from faeces to food. The main flies of importance are *Musca domestica*, the common housefly, and *Chrysomya putoria* and *C. megacephala*, the tropical green blowflies that are often found in latrine pits.

Secondly, culicine mosquitoes, particularly *Culex quinquefasciatus*, which is the major urban nuisance mosquito, breed preferentially in excreta-polluted waters such as wet pit latrines (see Chapter 3) or garbage-blocked stormwater drains. In many parts of the urban tropics they are the vector of Bancroftian filariasis, which is a disease caused by the nematode worm *Wuchereria bancrofti*. The adult worms live in the lymphatic ducts of humans, and embryo worms (called microfilariae) are shed in large numbers into the bloodstream at night. If a culicine mosquito ingests microfilariae during its blood meal, they develop inside the mosquito over a period of 10–15 days to become infective larvae. When the mosquito feeds again, they are introduced into another person where they develop over three to twelve months into adult male and female worms that establish themselves in the lymphatic system, and the cycle of microfilariae production starts again. After a few years of infection, damage to the lymphatic vessels occurs as a result of an immune response to the presence of the adult worms; lymph glands and lymphatic vessels become partially blocked and swollen as the lymph cannot drain. This leads to swellings of the genitalia, legs or arms, and the resulting gross deformity is called elephantiasis. Bancroftian filariasis is becoming increasingly common in urban areas that have good water supplies but poor sanitation—the resulting wastewaters pond in garbage-blocked stormwater drains and natural drainage channels, so permitting the culicine vector mosquitoes to proliferate.

Category VII: Excreta-related rodent-vector diseases

This category contains all Category I–III infections whose transmission can be facilitated by rodents, mainly rats, which carry the excreted pathogens on their bodies or in their intestinal tract from faeces to food.

It also contains leptospirosis, a disease caused by the bacterium *Leptospira interrogans* serovar *icterohaemorrhagiae*, which is primarily an infection of brown rats. Humans become infected when they come into contact with infected rat urine. Sewer workers are especially at risk, and so are those in contact with waters polluted by rat urine, such as stormwater channels and surface waters. Leptospirosis can also be due to different serovars that have different hosts: in the urban tropics serovars *caricola* and *pomona* are common and their hosts are dogs and pigs, respectively; human infection follows contact with urine-contaminated waters, as with serovar *icterohaemorrhagiae*. In humans the infection can be asymtomatic, mild (influenza-like symptoms) or severe: the severest form is Weil's disease, and this can be rapidly fatal if not treated; symptoms include jaundice, skin and eye haemorrhages, and liver and kidney failure.

2.3 EXCRETA-RELATED DISEASE CONTROL

The environmental classification of excreta-related diseases developed in subsection 2.2.2 is useful not only because it is informative in its own right, but (and more importantly for public health engineers) because it provides the framework for the control of these diseases through improved sanitation, and it indicates what else is required, in addition to sanitation, to achieve this (Table 2.2).

2.3.1 Control of Category I and II infections

These infections have a very direct person-to-person transmission. Improved personal and domestic hygiene, through improved water supplies, is extremely important, as is a sustained programme of hygiene education (see Chapter 14). Both improved water supplies—such that people have at least

Table 2.2 Control of excreta-related diseases

Category		Major control interventions
I.	Non-bacterial faeco-oral diseases	Improved water supplies Hygiene education Improved housing Provision of toilets
II.	Bacterial faeco-oral diseases	As for I, plus: Treatment of excreta or sewage prior to discharge or reuse
III.	Geohelminthiases	Provision of toilets Treatment of excreta or sewage prior to land application
IV.	Taeniases	As for III, plus: Meat inspection Thorough cooking of meat
V.	Water-based helminthiases	Provision of toilets Treatment of excreta or sewage prior to discharge or reuse Control of animal reservoirs Snail control Reduce water contact
VI.	Excreta-related insect-vector diseases	Domestic and peri-domestic hygiene Insect control
VII.	Excreta-related rodent-vector diseases	Peri-domestic hygiene Trapping Rodenticide application

25, and preferably closer to 50, litres of water per person per day—and hygiene education—so that they know how to use this amount of water properly to maximize the health benefits to them—are actually more important than sanitation in the case of Category I infections (sanitation is not, of course, unimportant, just less important).

With Category II infections, because of the greater persistence of the bacteria, sanitation becomes more important. Also, sewage requires effective treatment before it is reused in agriculture of aquaculture. "Effective treatment" means treatment specifically to kill the bacteria (see Chapter 10).

2.3.2 Control of Category III–V infections

Because of the extremely long persistence of these helminths, sanitation and hygiene education are very important for the control of these infections, more so than improved water supplies (although this is important for *Ascaris* control as its transmission route is often soil-to-mouth or soil-to-food-to-mouth). Effective sewage treatment is important as well, in order to prevent pasture, crop and water contamination.

The thorough cooking of meat, fish and aquatic vegetables is also necessary. Meat should be properly inspected in slaughterhouses.

Snail control can be achieved by the application of molluscicides, but these are expensive and usually have to be imported; and the snails quickly return when spraying ceases. The control of animal hosts is similarly difficult, but stray dogs should be controlled anyway to prevent rabies, and in endemic areas hygiene education programmes should include the importance of dog and pig excreta in the transmission of clonorchiasis and fasciolopsiasis.

2.3.3 Control of Category VI and VII infections

Fly control is best achieved by domestic (especially food) hygiene and by sanitation. VIP latrines (see Chapter 3) are excellent in controlling flies. Animal faeces are, however, a starting point for fly transmission of Category I and II diseases, and so need to be controlled as well as human faeces.

Mosquito control is more difficult, but still possible, with VIP latrines. Proper sullage disposal is also required (see Chapter 5) to avoid mosquito breeding. Stormwater drains should not be allowed to become blocked (commonly with garbage, so an efficient garbage collection service should be in place), or they will pond up and allow mosquitoes to breed.

Cockroaches are best controlled by homemade "boric acid flour balls". Mix two parts of boric acid crystals (H_3BO_3), two parts wheat flour and one part sugar into a light dough with milk; form into little balls and place wherever cockroaches are known to be (latrines, kitchens)—they readily consume the killer balls and quickly disappear. Sodium bicarbonate (baking powder) is an alternative to boric acid.

Rodent control is best achieved by scrupulous domestic and peri-domestic hygiene, supplemented by traps and rodenticides. Rats may be resistant to some commonly used rodenticides (anticoagulants, for example). Expert advice should be sought from local rodent control specialists.

2.4 FURTHER READING

A. S. Benenson (ed.), *Control of Communicable Diseases in Man*, 15th edition. American Public Health Association, Washington, DC (1990).

S. G. Berk and J. H. Gunderson, *A Color Atlas of Wastewater Organisms*. Lewis Publishers, Roca Baton, FL (1993).

M. Black, *Mega-slums: the Coming Sanitary Crisis*. WaterAid, London (1994).

D. Bradley, S. Cairncross, T. Harpham and C. Stephens, *A Review of Environmental Health Impacts in Developing Country Cities*. Urban Management Program Discussion Paper No. 6. The World Bank, Washington, DC (1991).

G. Cook (ed.), *Manson's Tropical Diseases*, 20th edition. W. B. Saunders Co. Ltd., London (1996).

C. F. Curtis (ed.), *Control of Disease Vectors in the Community*. Wolfe Publishing, London (1991).

G. Davidson, *Insecticides*. Ross Institute Bulletin No. 1. London School of Hygiene and Tropical Medicine, London (1988).

Environmental Health in Urban Development. Technical Report Series No. 807. World Health Organization, Geneva (1991).

S. A. Esrey, R. G. Feachem and J. M. Hughes, 'Interventions for the control of diarrhoeal diseases among young children: improving water supplies and excreta disposal facilities'. *Bulletin of the World Health Organization* **63**(4), 757–772 (1985).

R. G. Feachem, D. J. Bradley, H. Garelick and D. D. Mara, *Sanitation and Disease: Health Aspects of Excreta and Wastewater Management*. John Wiley & Sons, Chichester (1983).

D. Finer, *Parasites of Poverty*. Sarec (Swedish Agency for Research Cooperation with Developing Countries), Stockholm (1994).
A. Gray (ed.), *World Health and Disease*. Open University Press, Milton Keynes (1993).
J. E. Hardoy, S. Cairncross and D. Satterthwaite (eds), *The Poor Die Young: Housing and Health in Third World Cities*. Earthscan Publications, London (1990).
J. E. Hardoy and D. Satterthwaite (eds), *Squatter Citizen: Life in the Urban Third World*. Earthscan Publications, London (1989).
T. Harpham, T. Lusty and P. Vaughn (eds), *In the Shadow of the City: Community Health and the Urban Poor*. Oxford University Press, Oxford (1988).
T. Harpham and M. Tanner (eds), *Urban Health in Developing Countries: Progress and Prospects*. Earthscan Publications, London (1995).
P. J. Kolsky, "Diarrhoeal disease: current concepts and future challenges". *Transactions of the Royal Society of Tropical Medicine and Hygiene* **87** (S3), 43–46 (1993).
D. D. Mara and G. P. Alabaster, "An environmental classification of housing-related diseases in developing countries". *Journal of Tropical Medicine and Hygiene* **98** (1), 41–51 (1995).
W. Peters and H. M. Gilles, *A Colour Atlas of Tropical Medicine and Parasitology*, 3rd edition. Wolfe Medical Publications, London (1989).
R. J. Pleass, "Bites, bugs and bednets". *Biologist* **42** (1), 33–35 (1995).
J. B. Rose and C. P. Gerber, "Use of risk assessment for development of microbial standards". *Water Science and Technology* **24** (2), 29–34 (1991).
J. C. Sherris (ed.), *Medical Microbiology: An Introduction to Infectious Diseases*, 2nd edition. Elsevier, New York (1990).
I. Tabibzadeh, A. Rossi-Espagnet and R. Maxwell, *Spotlight on the Cities: Improving Urban Health in Developing Countries*. World Health Organization, Geneva (1989).
The Urban Health Crisis: Strategies for Health for All in the Face of Rapid Urbanization. World Health Organization, Geneva (1993).
Urbanization and its Implications for Child Health: Potential for Action. World Health Organization, Geneva (1988).

Further information on baited rat traps is available from:

Killgerm Chemicals Ltd
PO Box 2
Osset
West Yorkshire WF5 9NA
England

Fax: +44 1924 264757

3

Ventilated Improved Pit Latrines

3.1 DESCRIPTION

Ventilated improved pit latrines (Figure 3.1), now generally known as VIP latrines, are the modern version of the traditional pit latrine. Both odours and flies—the main problems with traditional pit latrines—are effectively eliminated by the action of the vent pipe (subsections 3.1.1 and 3.1.2), and the whole system is properly engineered as a permanent sanitation solution for urban areas by designing the pit either as a single pit, which is emptied mechanically, or as an alternating twin-pit system, which permits safe manual removal of the digested excreta (section 3.2).

VIP latrine pits receive excreta in the same way as any other pit latrine, by direct deposition through a squat hole (or a pedestal seat): urine infiltrates into the surrounding soil and the faecal solids are digested anaerobically. There is a slow accumulation of solids in the pit, which eventually requires emptying (section 3.2 and Chapter 7).

3.1.1 Odour control

The wind blowing over the top of the vent pipe creates a strong circulation of air through the superstructure, down through the squat hole, across the pit, and up and out of the vent pipe: any faecal odours from the pit contents are sucked up and exhausted out of the vent pipe, leaving the superstructure *entirely* odour-free. It was originally thought that this circulation of air was induced by the sun heating up the vent pipe and thus the air inside it, which became less dense and therefore rose (and this is why some vent pipes were painted

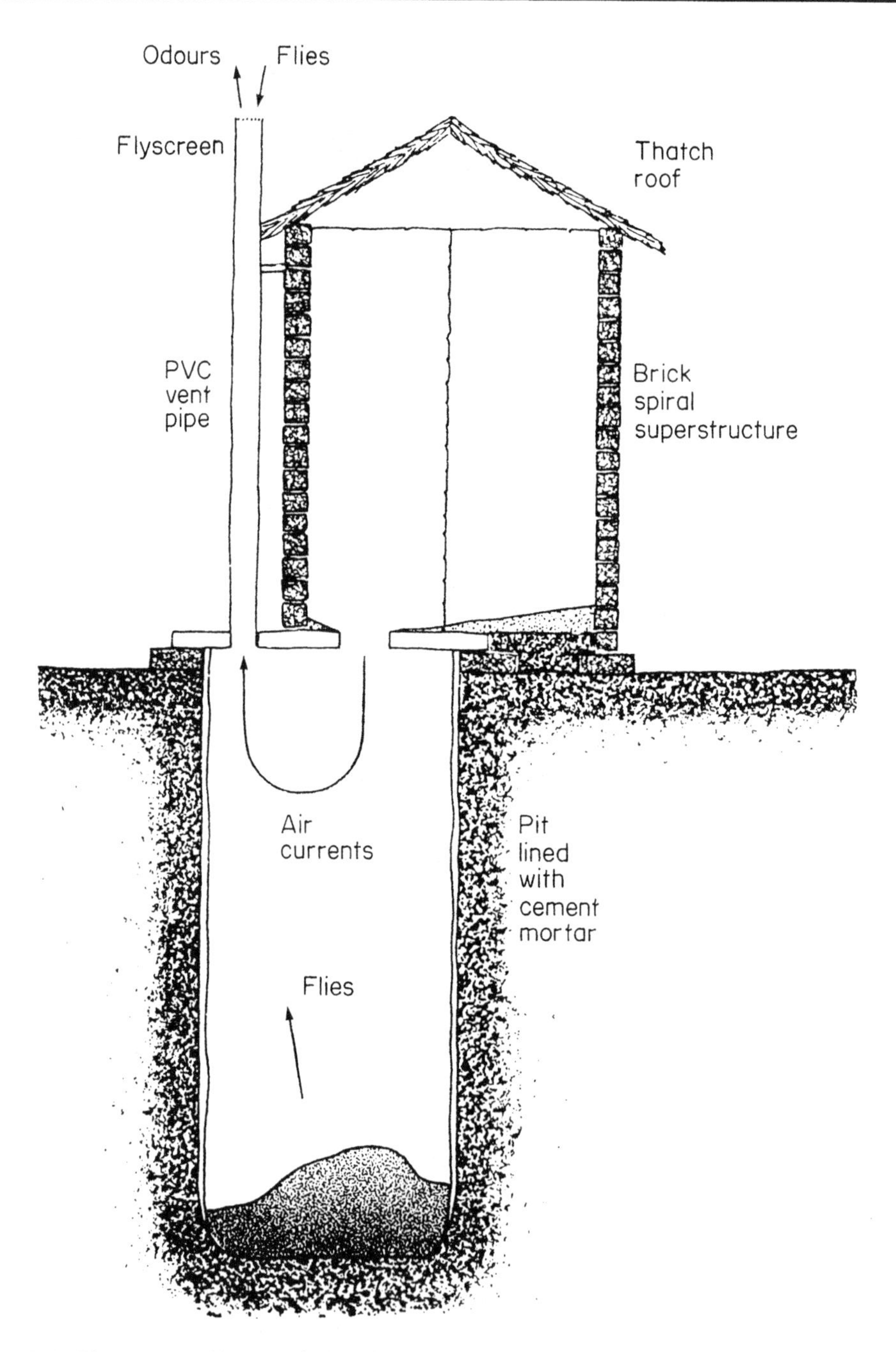

Figure 3.1 The ventilated improved pit latrine

black, to increase the absorption of solar heat). However, measurements in England and tropical Africa (Figures 3.2 and 3.3) have confirmed that the wind effect is more important

Figure 3.2 An urban VIP latrine in Botswana used to monitor wind speed, vent pipe air flow and temperatures

than that of the sun. Ventilation rates of over 20 m^3 of air per hour are readily obtained, and the VIP latrine really is odourless. (A good way to demonstrate the high circulation of air in VIP latrines is to hold a lighted newspaper near the squat hole: the smoke is sucked down through the squat hole and exits via the vent pipe for all to see.)

3.1.2 Fly and mosquito control

Flies are attracted to pit latrines by the faecal odours emanating from them, and gravid female flies know that faeces are a good place to lay their eggs. With VIP latrines, the odours come from the top of the vent pipe, and this is where the flies are attracted. However, the flies cannot enter as the top of the

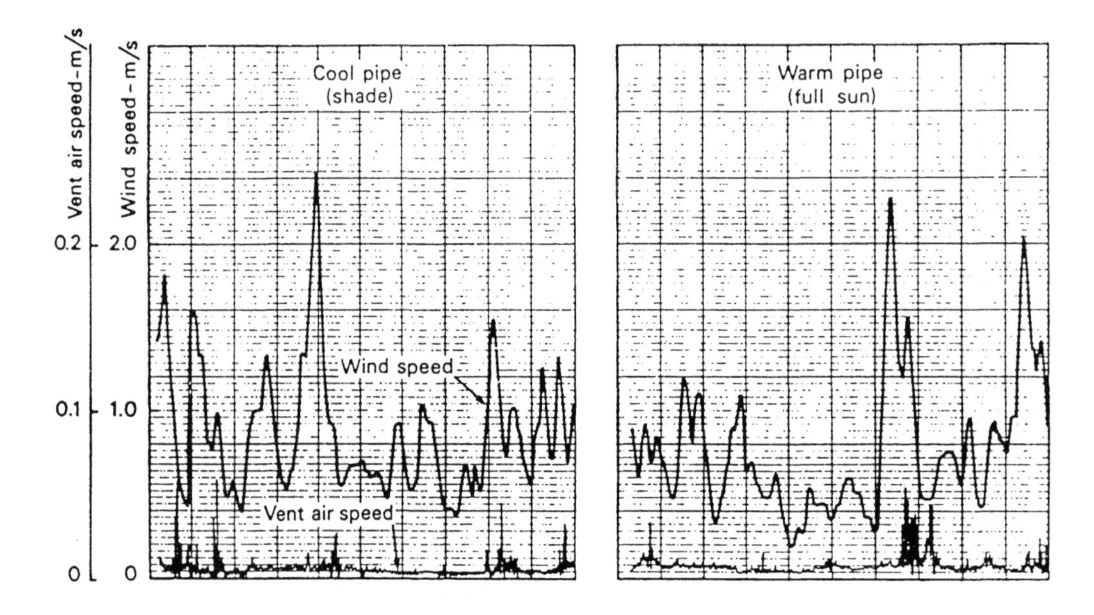

Figure 3.3 Recorder trace showing the coincidence of wind speed peaks and vent pipe air flow peaks

vent pipe is covered by a flyscreen (Figure 3.4). (Spiders soon learn that this is where the flies come, and it is common to see cobwebs at the top of vent pipes: Figure 3.5.)

However, one or two flies will enter the pit via the superstructure and squat hole, and lay their eggs. Eventually these eggs become newly emergent adults, and newly emergent adult flies instinctively fly in the direction of the strongest light (phototaxis). Provided the superstructure is reasonably well shaded, the strongest light visible to these newly emergent adults is the shaft of light coming down the vent pipe. So they fly up the vent pipe, but cannot leave because of the flyscreen. Unable to find any food, they eventually become exhausted and fall down into the pit, where they die. The vent pipe is extremely effective at fly control: in early experiments in Zimbabwe only 146 flies were caught leaving a VIP latrine over a period of 78 days, whereas the number leaving an unventilated, but otherwise identical, pit latrine over the same period was 13 953.

When the VIP latrine pit is a wet pit—that is to say, it penetrates the groundwater—the problem is not with flies, but rather with mosquitos. Unfortunately, newly emergent adult

Figure 3.4 Stainless steel flyscreen on the top of a VIP latrine vent pipe

mosquitos are not nearly as phototactic as are flies. So the vent pipe is of no use in mosquito control, and recourse has, therefore, to be made to other means. The best of these is the mosquito "lobster pot" trap shown in Figure 3.6; this is very effective and the users, once they see how many mosquitos are trapped and, correspondingly, how many fewer mosquito bites they have, quickly become very motivated to make sure the trap is always kept in place over the squat hole (of course, they are exposed to bites when using the latrine; again, people quickly learn by experience not to use the latrine during the period of maximum mosquito emergence, which is usually at dusk).

An alternative (or, on the belt and braces principle, an

Figure 3.5 Cobwebs at the top of a vent pipe—an example of sanitary ecology in action

additional) means of mosquito control is to add a cupful of kerosene to the pit each week. This floats above the water in the wet pit and suffocates the mosquito larvae (by lowering the surface tension so that they cannot cling to the underside of the surface and breathe through it). A more modern alternative to kerosene is expanded polystyrene beads: a 2 cm thick layer of 2 mm beads is required to stop mosquito breeding in wet pits (this is equivalent to around 30 l of beads per typical pit: the cost is some US$ 3). Polystyrene beads are very effective: their extremely high buoyancy ensures that they always return to the surface, and mosquito control is excellent: nightly catches of up to 13 000 *Culex quinquefasciatus* obtained before application of the beads have dropped to zero within two to three days.

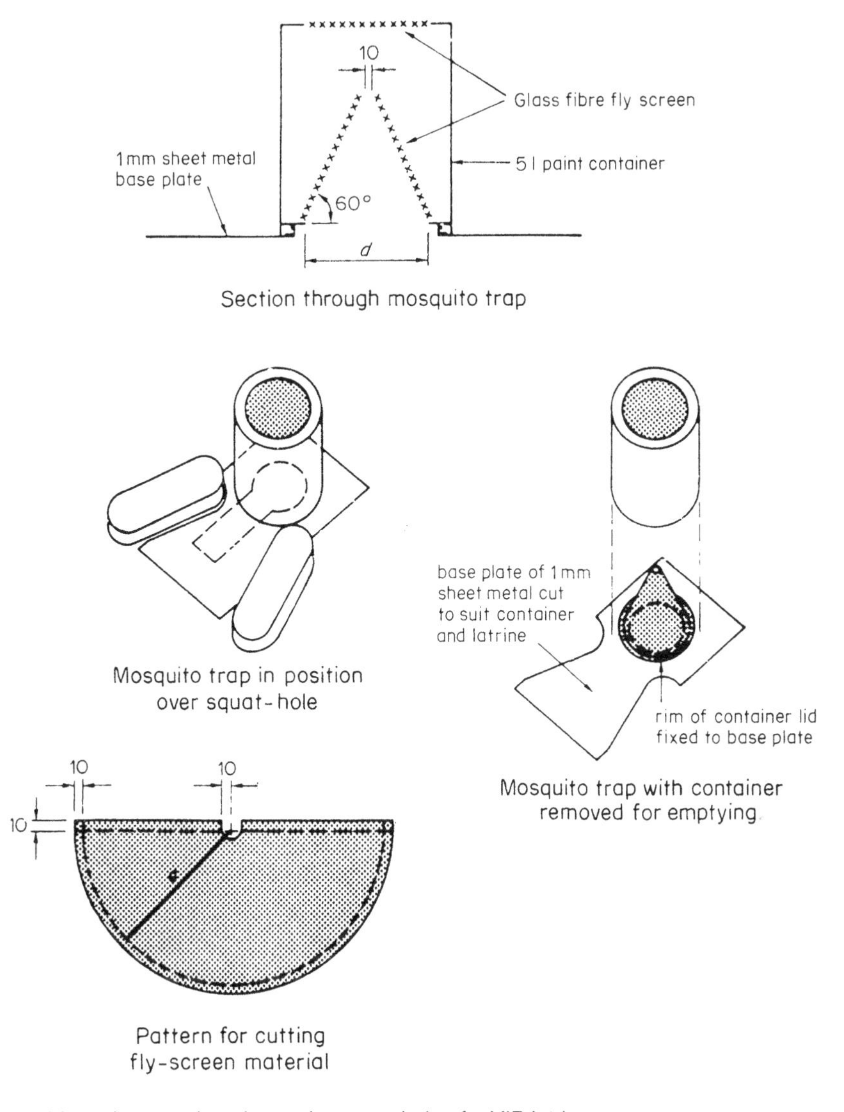

Figure 3.6 Mosquito trap placed over the squat-hole of a VIP latrine

3.1.3 Component parts

A VIP latrine comprises several parts:

- the pit, which is either a single pit or an alternating twin pit (subsection 3.2.2); in either case, the pit is lined in open-joint brickwork or blockwork—the lining prevents soil collapse during emptying operations or during heavy rains, and the open vertical joints permit liquids (urine) to infiltrate into the soil;

- a coverslab, usually of reinforced concrete, which covers the pit and has two holes—the squat-hole and one for the vent pipe;
- a superstructure and roof, for privacy and protection from rain and sun—it is best to build these in the same general architectural style as the house; and
- the vent pipe and flyscreen.

These are described in detail in sections 3.2–3.4.

3.2 THE LATRINE PIT

The latrine pit provides three functions: (a) digestion of fresh excreta, (b) storage of the digested excreta; and (c) infiltration of liquids into the surrounding soil. Excreta production depends on diet, but, in general, adults produce around 250 g of faeces and 1.2 l of urine per person per day. Digestion of faeces and infiltration of urine are both fairly rapid (the latter even in clay soils—see Table 4.1), so the pit is designed for the storage of digested solids. The effective pit volume (V, m^3) is given by:

$$V = rPn \tag{3.1}$$

where r = the solids accumulation rate (m^3 per person per year)
P = number of users of the latrine, both adults and children (= household size)
n = the interval between successive pit emptying operations (years)

Values of r are typically 0.03–0.06 m^3 per person per year in dry pits (those above the water table) and 0.02–0.04 in wet pits (those penetrating the water table). If bulky cleansing materials (such as corn cobs, cement bags) are used, these values should be increased by 50 percent.

In urban areas the value of n is two to five years, lower than the life of a VIP latrine in rural areas (10 years or so). A balance has to be struck between the costs of excavation and lining, which are governed by the pit volume, and the cost of emptying (section 7.5).

3.2.1 Single pits

Single-pit VIP latrines are used if the pits are to be emptied mechanically—manual emptying is not recommended as the excreta at the top of the pit are fresh and therefore may contain viable excreted pathogens, which would pose a risk to the emptiers. High-performance vacuum tankers are used to empty the pit; these are described in Chapter 7, which also discusses sludge disposal options. Access for desludging is provided by off-setting the pit from the superstructure, and covering the off-set part of the pit with removable coverslabs. The coverslab joint detail shown in Figure 3.7 is important: it prevents light entering the pit and so maintains fly control.

3.2.2 Alternating twin pits

Alternating twin-pit VIP latrines (Figure 3.8) are used if the pits are to be emptied manually (there is no point in emptying them mechanically: use single pits instead). Two pits are provided, each with its own vent pipe; the cover slab has two squat holes, one over each pit; and the superstructure is located centrally over the off-set pits.

Only one pit is used at a time, the squat hole over the other pit being closed by a concrete plug. The first pit is used for n years by which time it becomes full of digested excreta. It is then closed, and the second pit used for the next n years, when it too becomes full. The first pit is now emptied and put back into service. This alternating cycle continues indefinitely. Manual emptying is safe because the contents of each pit when it is emptied are between n and $2n$ years old and thus all the excreted pathogens, with the exception of a few *Ascaris lumbricoides* eggs (see subsection 2.2.2), will be dead. The value of n should never be less than one year to ensure pathogen die-off; usually $n = 2$ or 3 years, as this permits greater flexibility in emptying schedules: the pit can be safely

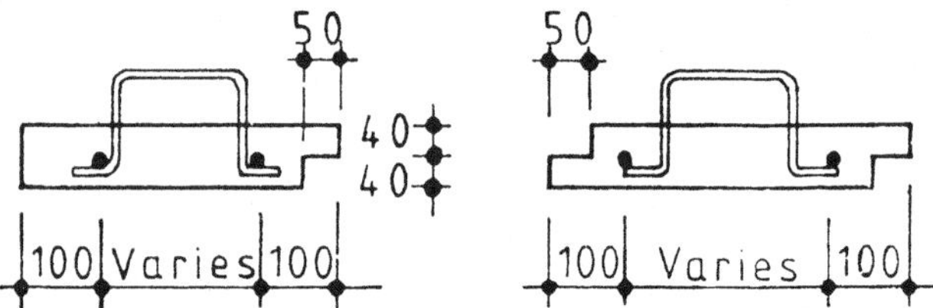

Figure 3.7 Removable joint detail: the overlap prevents light entering the pit

emptied in the three to six month period prior to it being put back into service.

This alternating sequence of pit usage, rest and emptying has to be carefully explained to the users, otherwise they might use both pits at the same time (this might seem extremely sensible to them—one pit for men and one for women, for example).

Manual emptying requires labour to perform this task. Is it available? This is a key question to answer prior to the implementation of alternating twin-pit technology. Perhaps the service is provided by the municipality, perhaps by a private firm contracted either by the municipality or directly by the householder; or maybe the householder is willing to empty his pit himself. These institutional aspects are very

Figure 3.8 Alternating twin-pit VIP latrine

important (see Chapter 15), as are the costs involved and exactly how the householder pays these costs (see Chapter 12).

Often householders are reluctant, at least initially, to empty their pits themselves; they may be totally against it: handling excreta is taboo in many societies. But once they see the pit contents immediately prior to emptying — that is, when they are $n - 2n$ years old (2–4 or 3–6 years old, usually) — their reluctance often disappears as the contents are not odorous faeces, but inoffensive composted material (much like garden soil) which has only a slight earthy, but definitely non-faecal, smell.

3.2.3 Wet pits

Alternating twin pits

If the groundwater table is less than around 1 m below ground level, the pits will be wet pits, and not generally suitable for suitable for manual emptying. Viable pathogens can travel from the wet pit in use to the other one, so re-contaminating its contents and thus posing a health risk to the emptiers.

Single pits

The problem with single pits is not their suitability in high groundwater table areas, but their excavation. If there is a seasonal variation in the position of the groundwater table, then obviously the pit should be excavated when the groundwater is at its lowest level.

Excavation under wet conditions can be difficult, but fortunately the pit depths of permanent, emptiable urban VIP latrines are not great, often only 1.5 m, although this clearly depends on the pit volume calculated from equation (3.1) (see section 3.8). Probably the best approach to the excavation of wet pits is to pump out the groundwater as excavation proceeds using a petrol-driven pump, and to insert short lengths of large-diameter concrete pipe as excavation proceeds. This 'mini-caisson' approach works well, but the concrete pipes must be made with sufficient holes to permit liquids to infiltrate into the soil.

High groundwater areas

If the groundwater table is within 30 cm of the ground surface, either permanently or seasonally, then the latrine coverslab should be raised by 30–50 cm above the ground level (Figure 3.9) to ensure that there is sufficient space for the air circulation required for odour control.

3.3 THE COVERSLAB AND FOUNDATION

The coverslab is best made in lightly reinforced concrete (for example, 75 mm thick with 6 mm bars at 100 mm centres). Squat-hole geometry is important: the holes are commonly either ovoid or keyhole-shaped, and they should not be too large or children might fear falling into the pit (Figure 3.10). Footrests can be provided on either side of the squat hole: these improve the depositional accuracy. The surface of the coverslab should slope gently towards the squat hole to permit

Figure 3.9 A single-pit VIP latrine raised above ground level in a high groundwater table area.

urine, and also any water used to clean the slab, to drain into the pit ('bucket' showers are sometimes taken in VIP latrines, and the sloping floor is helpful in this regard). A pedestal seat (Figure 3.11) can be provided if people prefer to sit during defecation (see Chapter 11).

The coverslab is set in cement mortar on a foundation of a single course of bricks laid on the ground in cement mortar abutting the pit lining. A fillet of mortar is then applied around the slab to prevent erosion of the foundation and pit lining.

It is important that the squat-hole (or pedestal seat) is *not* kept covered when the latrine is not in use. Covers prevent the circulation of air required for odour control. (Traditional unventilated pit latrines commonly rely on covers for fly control; if the users are familiar with these unimproved latrines, they may think a cover is also essential with VIP latrines—this is obviously an important point to stress in user education programmes: see Chapter 14.)

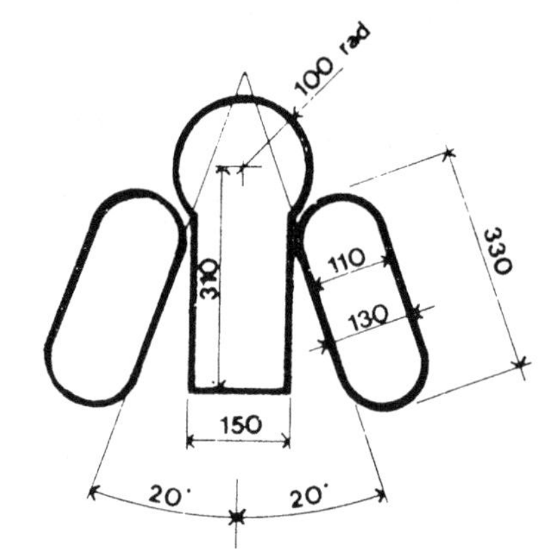

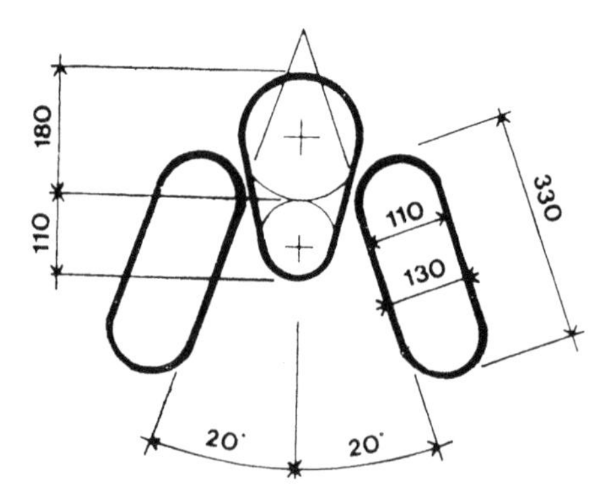

Figure 3.10 (a) Reinforced concrete coverslab and (b) squat-hole geometries

Figure 3.11 Pedestal seat unit for VIP latrines

3.4 THE SUPERSTRUCTURE AND VENT PIPE

The superstructure provides privacy and shelter. It is best made in the same building materials and style as the users' house—brickwork, blockwork or timber, for example, with a suitable roof in again whatever is used for the house—tiles, bituminous felt or thin ferrocement. Arguably the most important part of the superstructure is its entrance: the door should be connected to an internal counterweight so that it is self-closing (Figure 3.12); this is important because, if the door is left open when the latrine is not in use, newly emergent adult

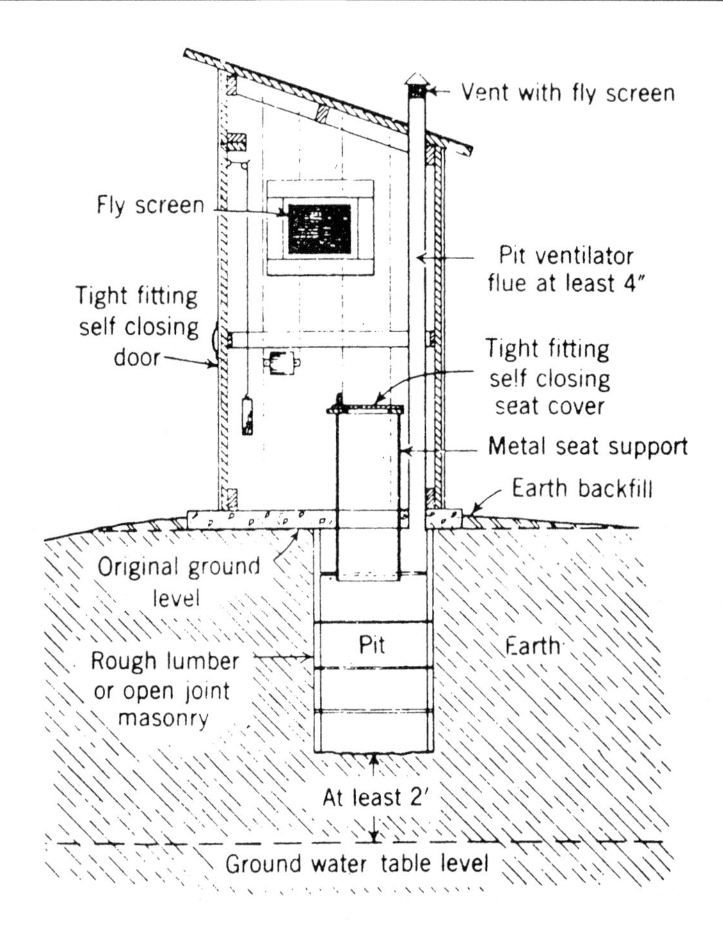

Figure 3.12 Early American VIP latrine showing self-closing door

flies are presented with an alternative source of strong light and so fly control ceases. Rural VIP latrines in Zimbabwe are usually designed with a spiral superstructure that keeps the interior sufficiently well shaded without the need for a door. An urban version of this is shown in Figure 3.13.

The vent pipe can be in 100 mm diameter PVC or made on-site in brickwork (225 mm square internally). The vent pipe should be completely straight, so as to allow light to enter the pit to ensure good fly control (subsection 3.1.2); and it should extend 500 mm above the roof to permit sufficient wind-induced air circulation for odour control (subsection 3.1.1). With a PVC pipe, care must be taken to ensure a fly-proof joint with the slab and to fix it securely to the superstructure. The flyscreen must also be fixed securely to the top of the vent pipe (Figure 3.4). The flyscreen apertures should not be greater than 1.2 × 1.5 mm to prevent flies and mosquitos passing through. The best material for the flyscreen

Figure 3.13 'Squared spiral' superstructure for a VIP latrine, which keeps the latrine sufficiently well shaded for fly control without the need for a door (although, of course, a door could be provided if the users so wished)

is stainless steel or aluminium; the latter is cheaper and lasts virtually indefinitely: the cost of 16/18 mesh screen made from 0.28 mm diameter aluminium wire is US$ 11 per m^2 ex works (see section 3.7) (the mesh number is the number of wires per inch, here 16 in one direction and 18 in the other).

3.4.1 In-house VIP latrines

It is possible to design VIP latrines so that they are in-house facilities (Figure 3.14); this may be preferred by some users. With in-house VIP latrines, the vent pipe should be 150 mm diameter, rather than 100 mm, so as to ensure good odour control.

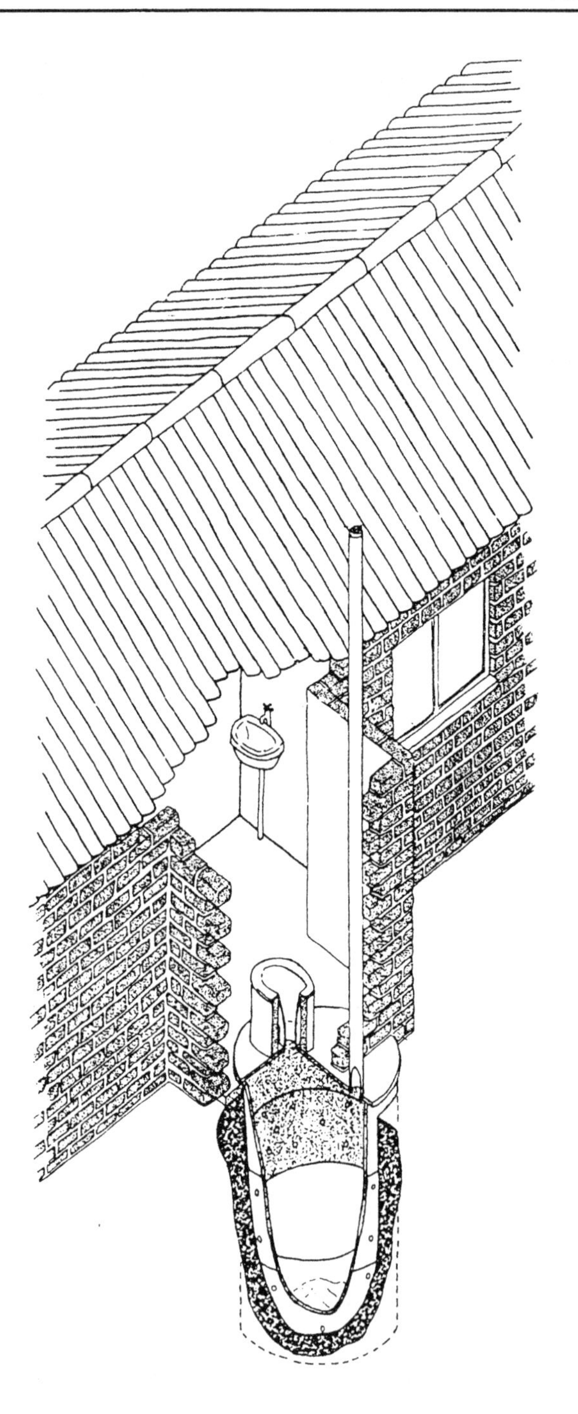

Figure 3.14 An in-house single-pit VIP latrine in northeast Brazil

3.5 OPERATION AND MAINTENANCE

Routine maintenance of urban VIP latrines is simply a matter of coverslab cleanliness and ensuring that the top of the vent pipe is free of leaves and anything else that might obstruct the airflow (Figure 3.15). Cockroach control is also usually required (see section 2.3). Intermittent visual inspection of the superstructure is a good idea, to detect any problems (for example, cracking) at an early stage. All this can be done by a member of the household—HLOM (see section 1.2).

Pit emptying is required every n (usually two or three) years. Either this is done by the householder or someone employed by him, in which case the householder has to remember when emptying is due (and it is easier to remember if n is small), or it is done by the municipality or a private firm contracted by the municipality (see Chapter 7).

3.6 GROUNDWATER POLLUTION

In many low-income urban areas water is not provided by a piped reticulation system, but, rather, obtained from open well or handpumps. It is, of course, possible that the groundwater may become polluted by bacterial and viral excreted

Figure 3.15 Fly-screen inspection using a mirror on the end of a pole

pathogens from VIP latrine pits. However, this is *not* necessarily an alarming situation because:

- groundwater is *not* sacrosanct: it is generally better to pollute the groundwater rather than have the pollutants on the ground surface—that is, it is better (in the sense that less disease will result) to have VIP latrines rather than no sanitation;
- groundwater pollution can be minimized by providing VIP latrine pits with a sand filter (Figure 3.16) or a sand envelope (Figure 4.10); and

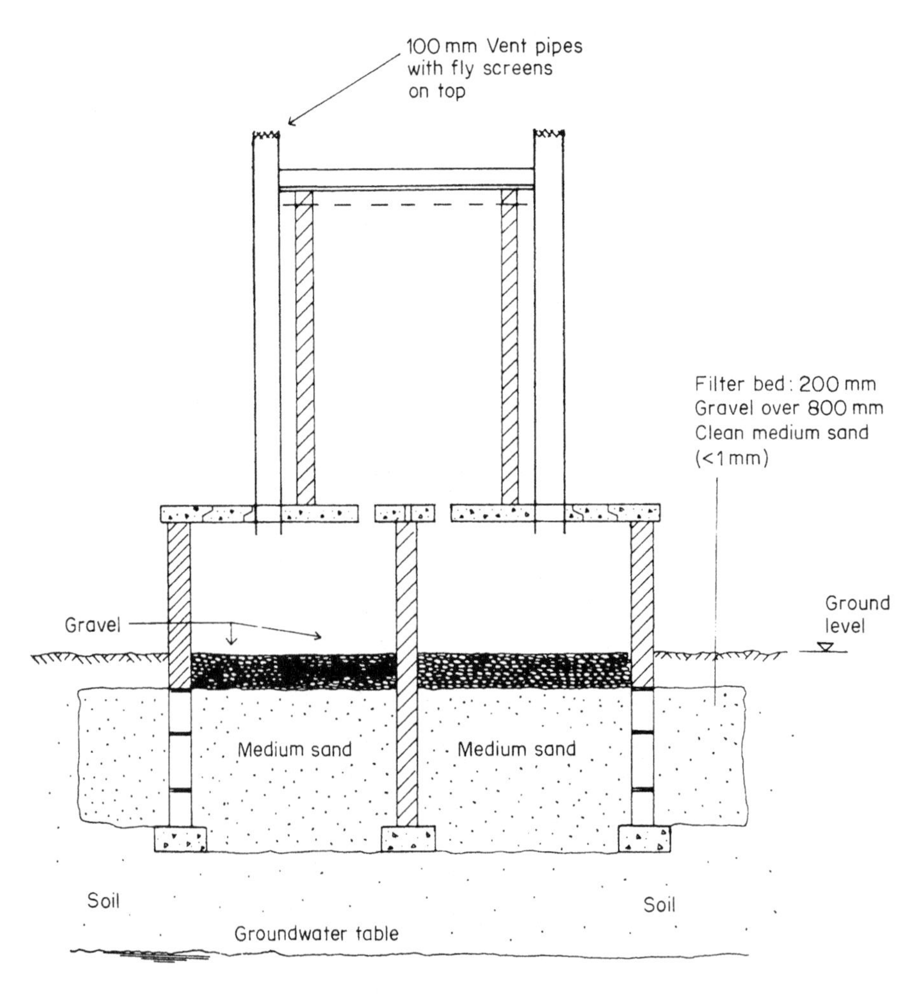

Figure 3.16 Raised alternating twin-pit VIP latrine with sand filter to minimize groundwater pollution

- off-site (reticulated) water plus on-site sanitation is cheaper than on-site water plus off-site sanitation (but see Chapter 9)—the off-site water can even come from the same aquifer but farther away from the latrine pits.

If there is at least 2 m between the pit base and the groundwater table, little microbial pollutant travel occurs in most unconsolidated soils. A horizontal distance between a well and a latrine of 10 m is often satisfactory; if this is not available, specialist hydrogeological advice should be sought. It should be remembered, however, that water supply improvements may be a more appropriate solution.

3.7 FURTHER READING AND INFORMATION

R. Brassington, *Field Hydrogeology*. John Wiley & Sons, Chichester (1988).

R. F. Carrol, *Sanitation for Developing Countries*. Overseas Building Note No. 189. Building Research Establishment, Watford (1982).

R. F. Carroll, *Health Aspects of Latrine Construction*. Overseas Building Note No. 196. Building Research Establishment, Watford (1991).

R. F. Carroll, *Disposal of Domestic Effluents to the Ground*. Overseas Building Note No. 195. Building Research Establishment, Watford (1991).

C. F. Curtis and P. M. Hawkins, "Entomological Studies of On-site Sanitation Systems in Botswana and Tanzania". *Transactions of the Royal Society of Tropical Medicine and Hygiene* **76**, 99–108 (1982).

W. J. Lewis, S. S. D. Foster and B. S. Draser, *The Risk of Groundwater Pollution by On-site Sanitation in Developing Countries*. Report No. 1/82. International Reference Centre for Wastes Disposal. Duebendorf (1982).

D. D. Mara, *The Design of Ventilated Improved Pit Latrines*. TAG Tchnical Note No. 13. The World Bank, Washington, DC (1984).

D. D. Mara, *Ventilated Improved Pit Latrines: Guidelines to the Selection of Design Options*. TAG Discussion Paper No. 4. The World Bank, Washington, DC (1985).

R. N. Middleton, "Making VIP latrines work". *Waterlines* **13** (4), 27–29 (1995).

P. R. Morgan and D. D. Mara, *Ventilated Improved Pit Latrines: Zimbabwean Brick Designs*. TAG Discussion Paper No. 1. The World Bank, Washington, DC (1985).

B. Ryan and D. D. Mara, *Pit Latrine Ventilation: Field Investigation Methodology*. TAG Technical Note No. 4. The World Bank, Washington, DC (1982).

B. Ryan and D. D. Mara, *Ventilated Improved Pit Latrines: Vent Pipe Design Guidelines*. TAG Technical Note No. 6. The World Bank, Washington, DC (1983).

J. van Nostrand and J. G. Wilson, *The Ventilated Improved Double-pit Latrine: A Construction Manual for Zimbabwe*. TAG Technical Note No. 3. The World Bank, Washington, DC (1983).
C. F. Ward, "Groundwater Quality Monitoring in Relation to On-site Sanitation in Developing Countries". *Journal of the Institution of Water and Environmental Management* **3** (3), 295–302 (1989).
C. Ward, "Groundwater quality monitoring in relation to on-site sanitation". *Waterlines* **8** (4), 11–14 (1990).

Aluminium flyscreens are manufactured by:

Cadisch Precision Meshes Ltd
Arcadia Avenue, Finchley
London N3 2JZ
England

Fax: +44 181 346 0613

VIP latrine building kits are available from:

Parry Associates Ltd
Overend Road
Cradley Heath
West Midlands B64 7DD
England

Fax: +44 1384 637783

3.8 DESIGN EXAMPLE

Design a VIP latrine for a family of 10. The groundwater table is at −2 m, and bulky anal cleansing materials are not used.

3.8.1 Single-pit VIP latrine

Assume, initially, that the latrine pit will be dry, so take $r = 0.05\ \mathrm{m}^3$ per person per year. Using equation (3.1) and assuming that the pit will be emptied every three years:

$$V = rPn$$
$$= 0.05 \times 10 \times 3 = 1.5\ \mathrm{m}^3$$

Suitable plan dimensions would be 1 m × 1.5 m, with a depth of 1.5 m (1 m effective depth plus 0.5 m free space). Thus, the pit base is above the groundwater table, and so the pit will be dry.

3.8.2 Alternating twin-pit VIP latrine

If the emptying cycle is three years as for the single-pit system, then each pit would be $1 \times 1.5 \times 1.5$ m, as above. A two-year cycle would result in smaller pits, for example $1 \times 1 \times 1.5$ m.

4

Pour-flush Toilets

4.1 DESCRIPTION

The pour-flush (PF) toilet (Figure 4.1) has three main parts: (*a*) the superstructure, which houses (*b*) the latrine pan with its integral waterseal, which is connected by small diameter pipework to (*c*) single or alternating twin leach pits. Instead of having a separate superstructure, the PF toilet can be located in-house. Odour and insect control is achieved by the waterseal (as in conventional cistern-flush toilets).

After the toilet is used, it is manually pour-flushed with 2–3 l of water. The excreta are flushed through the pan and trap, along the pipework into the leach pit (Figure 4.2). Some of the clean flushwater remains in the trap, so maintaining the waterseal (and hence odour and insect control). Around 5–10 litres per caput per day (lcd) of wastewater (excreta and flush water) enter the pit, together with an additional usually equal amount if water is used for anal cleansing. The leach pit has to provide sufficient volume for solids storage (as for VIP latrines—section 3.2), as well as sufficient area for the wastewater to infiltrate into the soil (section 4.2.). If the soil is unsuitable for infiltration, the liquid effluent can be removed by settled sewerage (Chapter 8).

Single leach pits are appropriate in urban areas only if they can be emptied mechanically (Chapter 7), since their contents are not pathogen-free. Alternating twin pits (see subsection 3.2.2) must be used if the pits are to be emptied manually, as the rest period ensures that the material to be removed is essentially free of excreted pathogens (only a few eggs of *Ascaris lumbricoides* will be viable after one to two years). With twin pits, a flow-diversion box (Figure 4.3) is required, only one exit of which is open at any one time, the other being closed off with, for example, a brick and a piece of sacking material.

The latrine pan and waterseal can be a squat-pan or

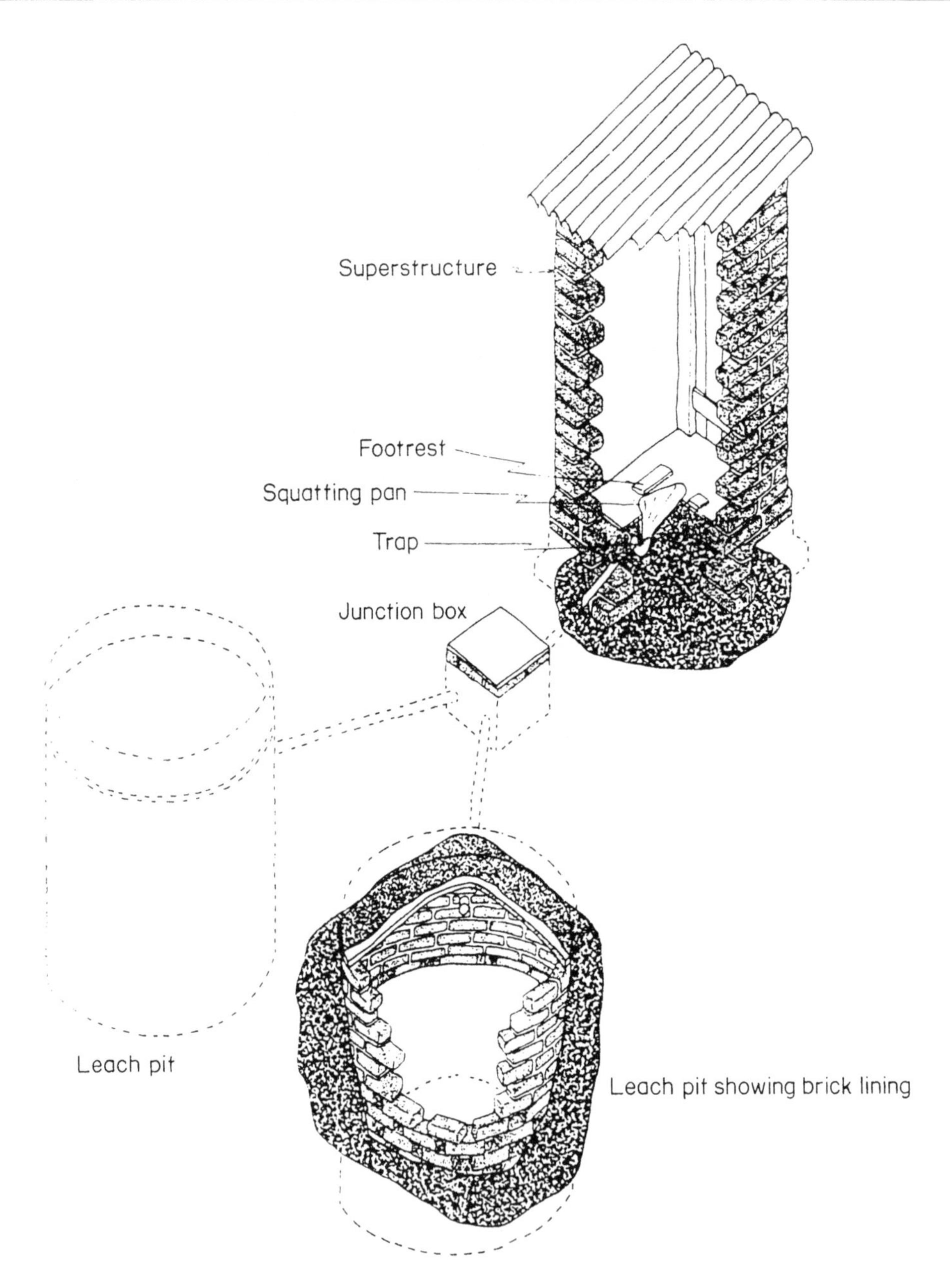

Figure 4.1 Schematic drawing of a pour-flush toilet with alternating twin leach pits

pedestal unit, depending on whether the users prefer to squat or sit (Chapter 11). Figure 4.4 shows the glass-fibre-reinforced plastic squat-pan and trap used in India; its ergonomic dimensions allow efficient flushing of the excreta, prevent urine splashing, and permit easy use and comfortable squatting by

Figure 4.2 Excreta being flushed from PF toilet into the leach pit

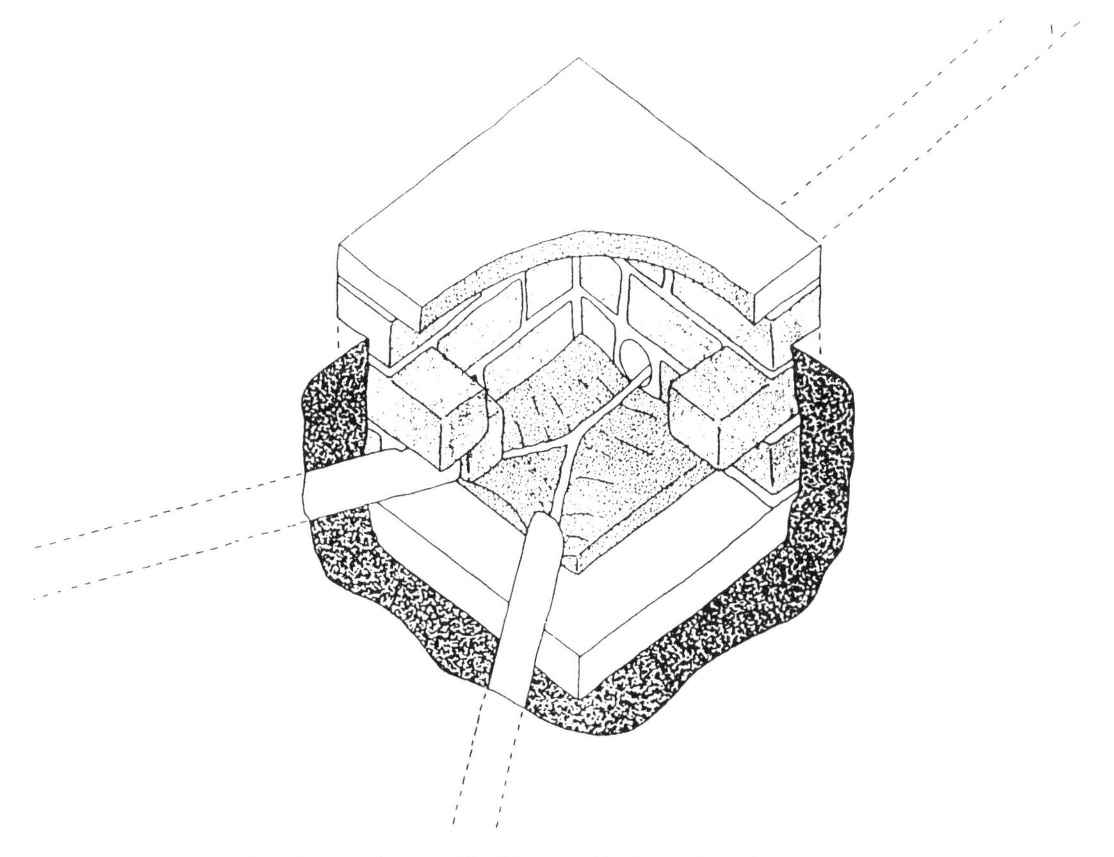

Figure 4.3 Flow-diversion box for use with PF toilet with alternating twin pits

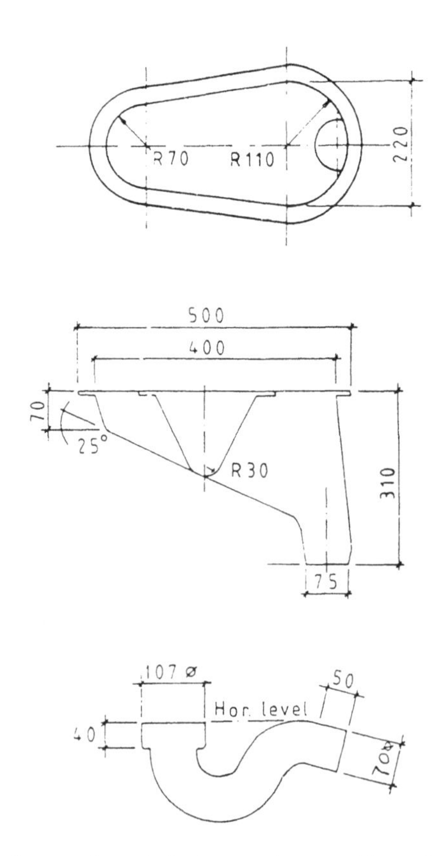

Figure 4.4 (a) Indian glass-fibre-reinforced plastic squat pan and trap. (b) dimensions (mm)

both children and adults. The unit can be made in various materials, such as high-density polyethylene, ceramics or sand–cement mortar. A ceramic pedestal unit, for those who prefer to sit, is shown in Figure 4.5.

The depth of the waterseal in the trap unit (or within the pedestal unit) is critical: if it is too great, the volume of flushwater required is too high; if it is too small, construction inaccuracies might mean that a waterseal is not formed, with the result that odours and insects are not controlled (Figure 4.6). A depth of 20–30 mm is optimal. The diameter of the pipe connecting the pan and trap (or pedestal unit) to the leach pit should be 75–100 mm, and the distance between them no more than 10 m.

4.1.1 Conversion to cistern-flush operation

Squat-pan and pedestal units can be simply upgraded, if the users so wish, to operate as low-volume cistern-flush toilets.

Figure 4.5 Brazilian ceramic PF pedestal seat unit

Figure 4.6 Checking the squat pan level during installation

The Indian squat-pan unit shown in Figure 4.4 is converted by installing the add-on unit shown in Figure 4.7. In this mode of operation the flush volume is only 1.5 l; the cistern, which has a capacity of 15 l (although this could be more), is actually in two interconnected parts: when the toilet is flushed, the outlet from the lower part discharges the 1.5 l flush but the outlet from the upper part to the lower part is closed; when the flush handle is released, the lower outlet closes and the upper outlet opens, so refilling the lower compartment in readiness for the next flush. The cistern can be filled via a ballcock from the in-house water supply, or manually; it is thus especially suitable for use in urban areas that have an intermittent water supply (i.e. 1–2 h in the morning, and possibly also in the afternoon or evening).

The Brazilian PF pedestal seat unit shown in Figure 4.5 is simply upgraded by installing a low-volume cistern connected to the in-house water supply and a down pipe (Figure 4.8). The flush volume in cistern-flush mode is 5 l.

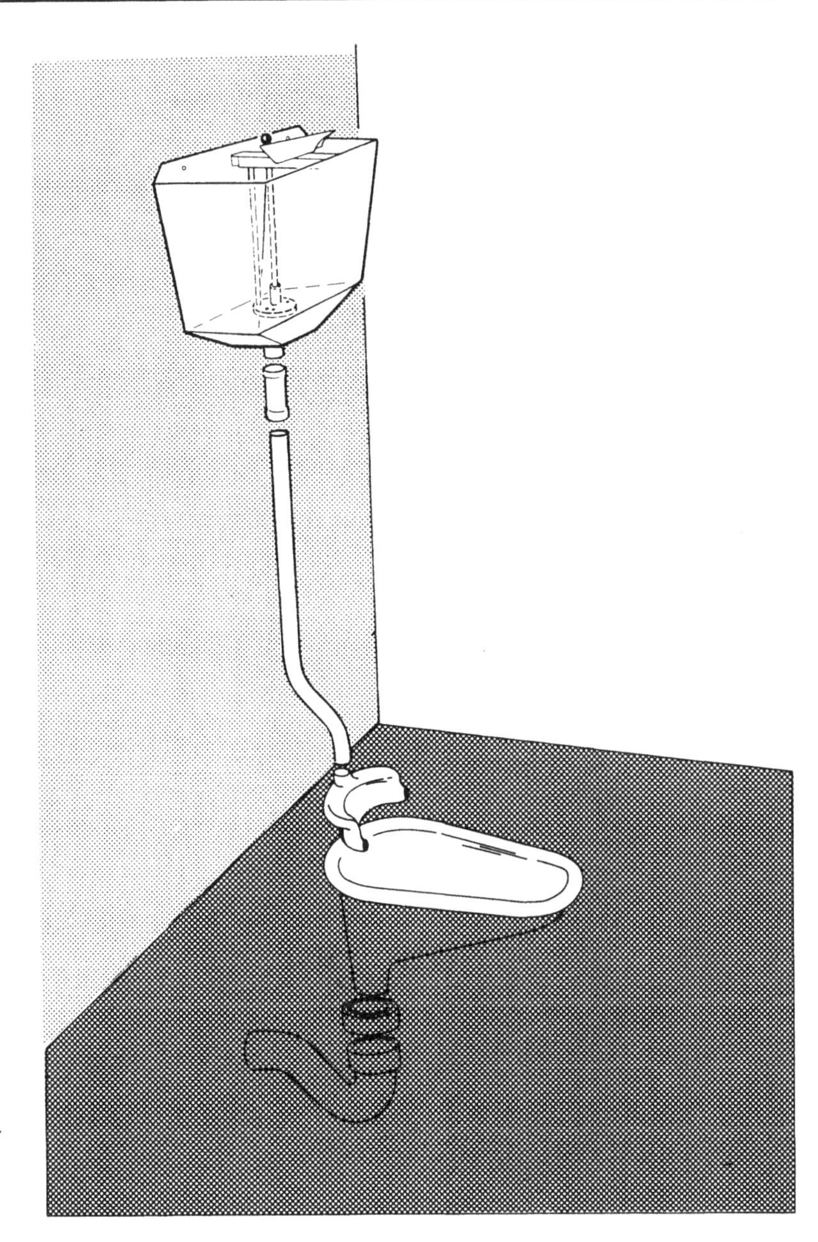

Figure 4.7 Add-on system to convert Indian PF system to a low-volume cistern-flush operation

4.1.2 Advantages

The PF toilet is a well-tried, robust technology that is in widespread use in developing countries. Its principal merits are:

- low capital costs (for example, in India it costs around US$ 200, but this can be reduced by self-help labour);

Figure 4.8 Brazilian PF system converted to a low-volume cistern-flush operation

- low water requirement, under 10 lcd for flushing;
- high social acceptability, especially where water is used for anal cleansing (but it is also popular elsewhere; in Latin America, for example);
- HLOM requirements are minimal—the pan (or pedestal) unit and the floor must be kept clean;
- odour and insect control is excellent;
- easy and safe for use by children;
- easy upgrading (see Chapter 13).

PF toilets would be unsuitable if bulky anal cleansing materials are used, if a minimum of 5 lcd of water for flushing cannot be relied on, or if incomes are extremely low and subsidies are unavailable (see Chapter 15).

4.2 LEACH PIT DESIGN

Leach pits for PF toilets have to be designed for solids storage and for infiltration. The former is done in exactly the same way as for the VIP latrine pits (equation (3.1)). The latter is necessary to ensure that the greater volume of wastewater (flush water, urine and any water used for anal cleansing) can pass into the surrounding soil. Infiltration refers to the passage of this wastewater across the pit-soil interface, which is partially covered in a bacterial slime layer; once across, the wastewater then percolates through the soil. This is the reason why percolation tests, which measure how quickly clean water passes through undisturbed soil, should not be used to measure infiltration. The long-term infiltration rate depends on the type of soil; suitable design values are given in Table 4.1. Alternatively, the long-term infiltration rate (I, l/m^2 day) may be estimated from the *in situ* soil conductivity (k, m/s) (determined by a suitable standard soil mechanics test procedure) by the following empirical equation, which takes blocking of the pit–soil interface into account:

$$I = 4k \times 10^4 - [49/\log_{10}(197k)] \qquad (4.1)$$

The pit sidewall area required for infiltration (A_i, m^2) depends in the wastewater flow (Q, l/day) and the long-term infiltration rate:

$$A_i = Q/I \qquad (4.2)$$

The wastewater flow depends on the number of users, how frequently the toilet is flushed (is it flushed after every use or only after each defecation?), the flush volume (litres per flush), urine volumes, and whether water is used for anal cleansing. Generally the flow is 5–20 lcd.

Table 4.1 Design values for long-term infiltration rates for domestic wastewater into various types of soil

Soil type	Long-term infiltration rate (l/m^2 day)
Sand	50
Sandy loam	30
Porous silty loam, porous silty clay loam	20
Compact silty loam, clay[a]	10

[a]Expansive clays should be absent as they are not suitable for infiltration.

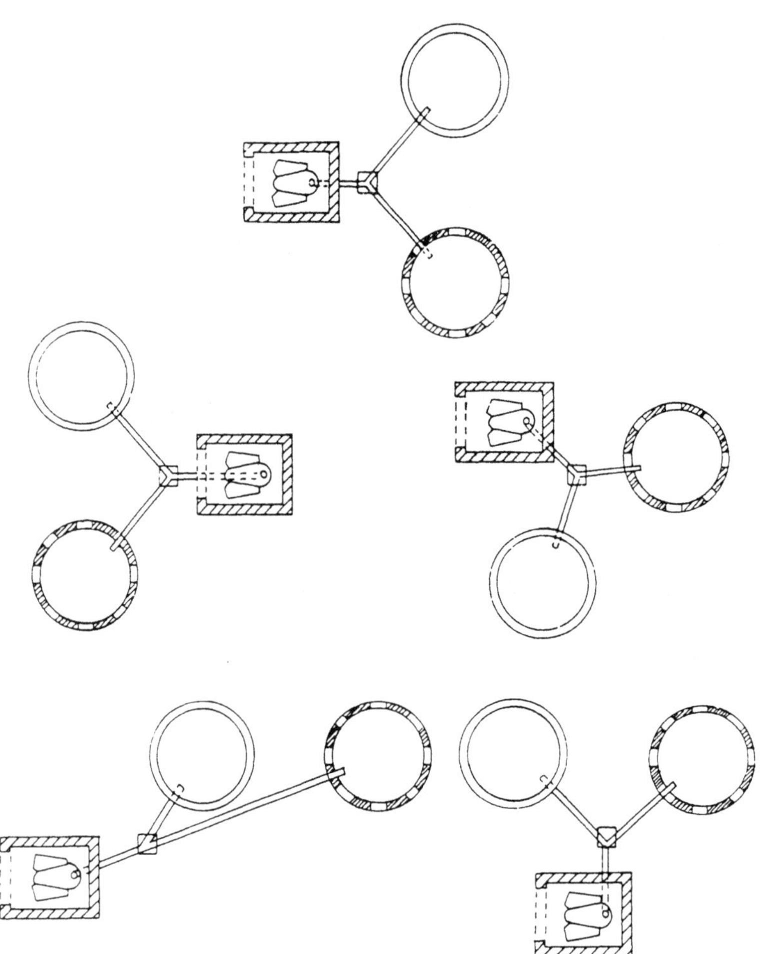

Figure 4.9 Alternating twin leach pits can be located in any one of a variety of relative positions

The pit volume (V_i, m^3) corresponding to this sidewall area is now calculated; for example, for a circular pit of diameter D, m:

$$V_i = \pi D^2 h/4 \qquad (4.3)$$

where h = the height of the sidewall area = $A_i/\pi D$. So:

$$V_i = A_i D/4 \qquad (4.4)$$

i.e.

$$V_i = QD/4I \qquad (4.5)$$

For alternating twin leach pits, the effective volume of each pit is either V_s (equation (3.1)) or V_i, whichever is greater.

For single leach pits, the effective volume (V, m^3) is given by:

$$V = V_s + V_i \tag{4.6}$$

This is slightly conservative as, of course, some infiltration occurs through the sidewall area corresponding to V_s. But this better permits restoration of the infiltrative capacity after emptying (with alternating twin pits this occurs during the rest period).

PF leach pits should always be lined in brickwork (Figure 4.2) or blockwork in order to prevent sidewall erosion due to the relatively large volumes of wastewater entering the pit. Alternating twin leach pits should be separated by a minimum of 1 m; a variety of relative positions is possible (Figure 4.9).

4.2.1 Prevention of groundwater pollution

If it is desired to minimize groundwater pollution (but see section 3.6), the pit base should be sealed with lean concrete and a 0.5 m annulus of sand placed between the pit lining and the soil (Figure 4.10). The effective size of the sand should be < 1 mm. Faecal bacterial travel is considerably reduced, that

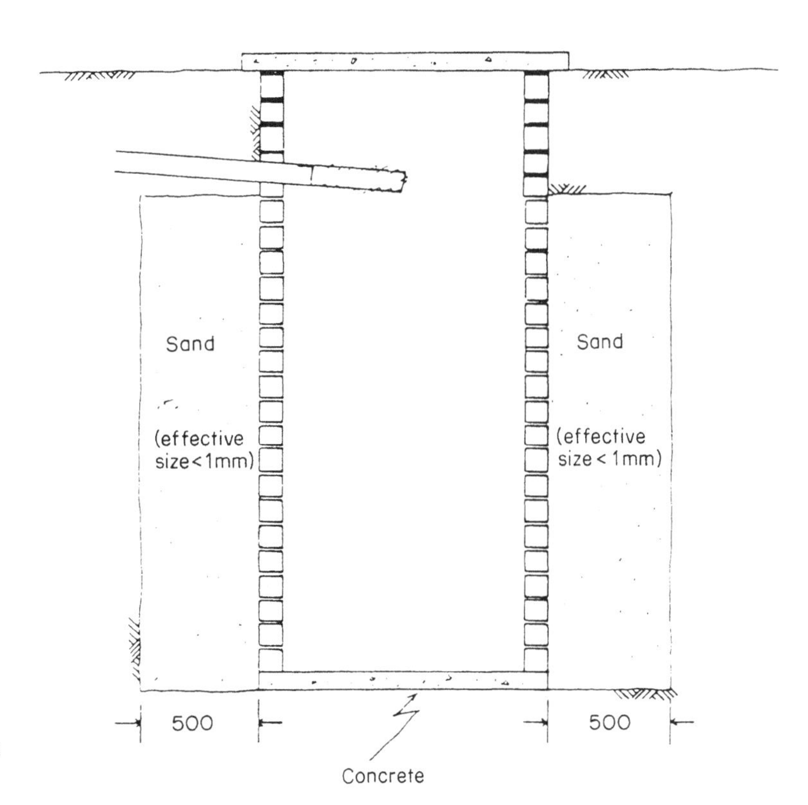

Figure 4.10 Sand envelope around leach pit lining for groundwater pollution control

of helminth eggs and protozoan cysts is effectively prevented, and excreted viruses become adsorbed onto the sand particles.

4.3 FURTHER READING AND INFORMATION

D. D. Mara, *The Design of Pour-flush Latrines*. TAG Technical Note No. 15. The World Bank, Washington DC (1985).

E. Olsson, *Low-volume WC Systems: A Development Project in India*. Research Report No. TN4. National Swedish Institute for Building Research, Gävle (1988).

A. K. Roy, *Manual on the Design, Construction and Maintenance of Low-cost Pour-flush Waterseal Latrines in India*. TAG Technical Note No. 10. The World Bank, Washington DC (1984).

Technology Advisory Group—India, *Report of the Committee on Design Criteria for Pour-flush Waterseal Latrines for Rural Communities in India*. TAG-India, New Delhi (1985).

The Indian PF squat-pan and trap are manufactured by:

Indian Institute of Low-Cost Sanitation
B-1/15, Community Centre
Janakprui
New Delhi 110058
India

The conversion kit for upgrading the Indian PF squat-pan to cistern-flush mode is manufactured by:

E. I. D. Parry (India) Ltd
Dare House
234 NSC Bose Road
Madras 600001
India

The Brazilian PF pedestal seat unit (called a *vaso de descarga reduzida*, or VDR) is manufactured by:

Duratex SA
Avenida Paulista 7
Bela Vista
01311-900 São Paulo—SP
Brazil

Fax: +55 11 884 8133

4.4 DESIGN EXAMPLE

Design the leach pit (single and alternating twin) for a PF toilet serving a family of eight. Wastewater flow is 15 lcd and the soil is a porous silty loam.

4.4.1 Infiltration

Use equation (4.2) with $I = 20\ \mathrm{l/m^2}$ per day (Table 4.1):

$$A_i = Q/I$$
$$= 8 \times 15/20 = 6\ \mathrm{m^2}$$

Use equation (4.4) for a circular pit of diameter 1.2 m:

$$V_i = A_i D/4$$
$$= 6 \times 1.2/4 = 1.8\ \mathrm{m^3}$$

4.4.2 Solids storage

Use equation (3.1) with $r = 0.04\ \mathrm{m^3}$ per person year and $n = 3$ years:

$$V_s = rPn$$
$$= 0.04 \times 8 \times 3 = 1\ \mathrm{m^3}$$

Single pit system

The effective volume V is $(V_i + V_s) = 2.8\ \mathrm{m^3}$. The pit dimensions are: 1.2 m diameter, 3 m deep (the depth being $V/(\pi D^2/4)$ plus 0.5 m). Alternative dimensions could be, for example, 1.5 m diameter, 2.1 m deep.

Alternating twin-pit system

V_i is greater than V_s, so the effective volume of each pit is $1.8\ \mathrm{m^3}$. Suitable dimensions would be: 1.2 m diameter, 2.1 m deep.

5

Sullage Disposal

5.1 SULLAGE VOLUMES

Sullage is all household wastewater except wastewater from toilets; it is sometimes called greywater (and toilet wastewaters are termed blackwaters). The volume of sullage produced depends on the type of water supply. Urban households relying on hand-carried water supplies (from public taps or wells, for example) consume around 20–30 litres per caput per day (lcd), although this depends on where clothes are washed: if this is done at a public clothes-washing facility, rather than

Figure 5.1 A stream of sullage in a West African city

at home, the at-home water consumption is less, perhaps around 10–20 lcd. With yard-tap water supplies, water consumption is much higher, around 40–80 lcd. The volume of sullage produced is around 80–90 percent of the at-home water consumption.

Sullage needs to be disposed of hygienically as it is only slightly less hazardous than other wastewater, and it can encourage mosquito breeding with the attendant risk of Bancroftian filariasis (section 2.2).

5.2 SULLAGE DISPOSAL ALTERNATIVES

Households with VIP latrines or PF toilets need to dispose of their sullage separately as these systems are not designed for sullage disposal. Often it is simply discharged into the alleyway adjacent to the house, but this is clearly undesirable: the

Figure 5.2 A sullage drain in Olinda, northeast Brazil

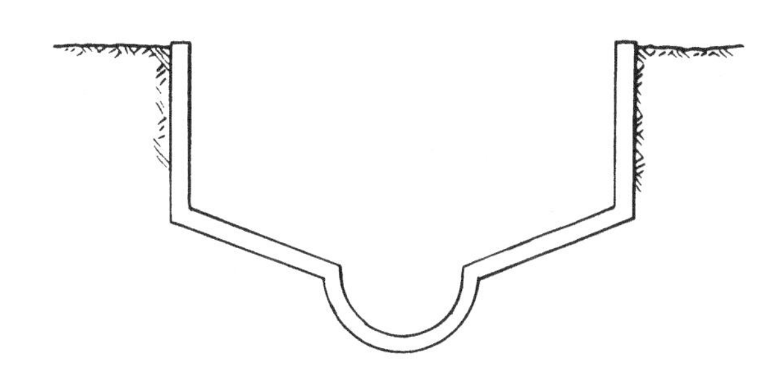

Figure 5.3 Section of combined sullage–stormwater drain: the circular section carries the sullage flow in dry weather periods

result is a fairly steady stream of sullage (Figure 5.1), which eventually joins a natural drainage channel.

A more appropriate solution is to discharge the sullage into a hydraulically well-designed stormwater drain (Figures 5.2 and 5.3), or into a sullage soakaway. The soakaway can be a simple 1–2 m deep pit filled with rocks; it should be sized in the same way as a PF toilet leach pit (section 4.2, but use only the volume for infiltration (V_i) and, since sullage infiltrates at least twice as fast as toilet wastewater, use values for the long-term infiltration rate that are twice those given in Table 4.1).

5.3 FURTHER READING

J. H. T. Winneberger (ed.), *Manual of Greywater Treatment Practices*. Ann Arbor Science, Ann Arbor, MI (1974).

6

Septic Tanks

6.1 DESCRIPTION

Septic tanks are small rectangular chambers, usually sited just below ground level, in which household wastewater (toilet wastewater and sullage) is retained for 1–3 days. Most commonly they are constructed in brickwork or blockwork and rendered internally with cement mortar to ensure watertightness. During this time the solids settle to the bottom of the tank, where they are digested anaerobically. A thick crust of scum is formed at the surface and this helps to maintain anaerobic conditions. Although digestion of the settled solids is reasonably good, some sludge accumulates and the tank must be desludged at regular intervals, usually once every one to five years (section 6.4). The effluent from septic tanks is disposed of either on-site (section 6.5) or taken off-site by settled sewerage (Chapter 8). Although septic tanks are most commonly used to treat the sewage from individual households, they can be used as a communal facility for populations up to about 300.

6.2 APPROPRIATENESS

Septic tanks are most appropriate in low- to medium-density urban areas. They are generally not an option for low-income areas, for which the lower cost options described in Chapters 3–5 and 9 are more suitable. Nonetheless, septic tanks either with on-site disposal of their effluent or off-site disposal by settled sewerage are normally less expensive than conventional sewage, so serving middle and upper income areas with septic tanks leaves (or should leave) more resources available to serve low-income areas.

6.3 TANK DESIGN

The septic tank design procedure presented here is based on the Brazilian septic tank code, which takes a more rational approach to design than others. The tank is considered to be made up of four zones, each of which serves a different function (Figure 6.1):

- scum storage zone
- sedimentation zone
- sludge digestion zone, and
- digested sludge storage zone.

6.3.1 Scum storage

Scum accumulates at approximately 30–40 percent of the rate at which sludge accumulates and so the tank volume for scum storage (V_{sc}, m^3) can be taken as 0.4 V_{sl} (see subsection 6.3.4).

6.3.2 Sedimentation

The time required to permit sedimentation of settleable solids decreases with the number of people served (this reflects the reduced peak flows that occur as the population served increases), according to the equation:

$$t_h = 1.5 - 0.3 \log(Pq) \tag{6.1}$$

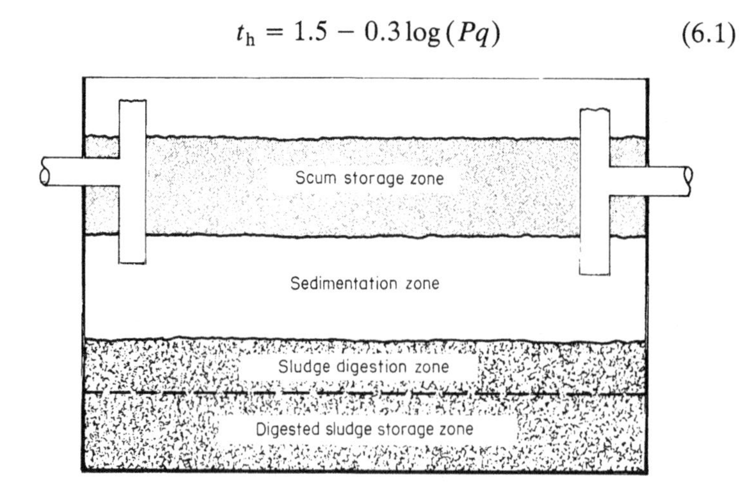

Figure 6.1 The four different functional zones in a septic tank.

where t_h = minimum mean hydraulic retention time for sedimentation, days
P = contributing population
q = wastewater flow per person, l/day.

The value of t_h used should not be less than 0.2 day. The tank volume for sedimentation (V_h, m^3) is given by:

$$V_h = 10^{-3} P q t_h \tag{6.2}$$

6.3.3 Digestion

The time needed for the anaerobic digestion of the settled solids (t_d, days) varies with temperature (T, °C), as shown in Figure 6.2 and given by the equation:

$$t_d = 1853 T^{-1.25} \tag{6.3}$$

An alternative way of obtaining a value for t_d is to consider

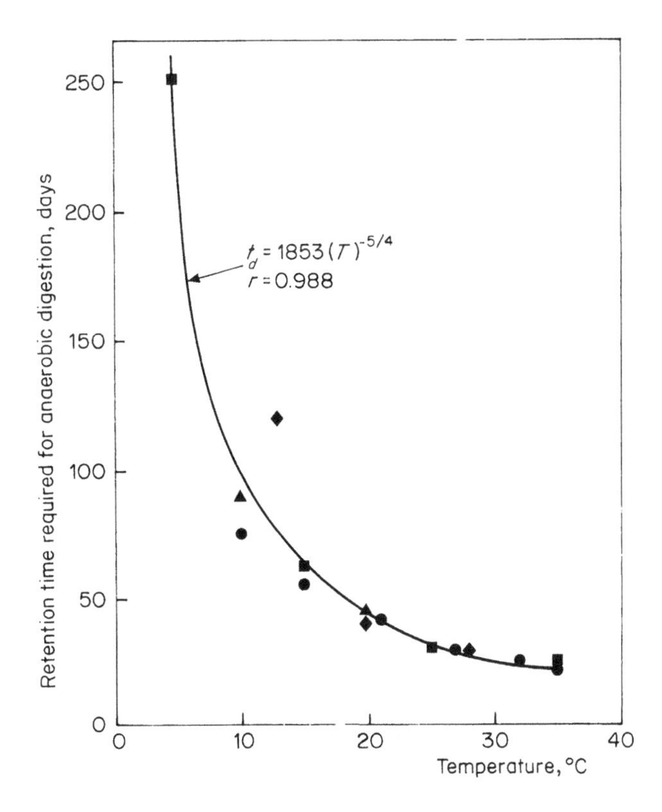

Figure 6.2 Sludge digestion times at various temperatures.

the process growth kinetics of a completely mixed anaerobic digester. The minimum retention time for the anaerobic biomass (θ_{MIN}, days) depends on how fast the bacteria can utilize their food supply, and on how much food is needed to generate additional biomass. If "food" is taken to be organic matter, or biochemical oxygen demand (BOD—see Chapter 10), then:

$$\theta_{MIN} = 1/Yk_T \quad (6.4)$$

where Y = yield coefficient, mg cells produced per mg BOD utilized
k_T = maximum BOD utilization rate, per day.

The value of Y is around 0.04 for high lipid wastes. In the temperature range 20–35 °C, k_T varies with temperature as follows:

$$k_T = 6.67\ (1.035)^{T-35} \quad (6.5)$$

Combining equation (6.4) and (6.5) with $Y = 0.04$, gives:

$$\theta_{MIN} = 3.75\ (1.035)^{35-T} \quad (6.6)$$

A fairly large factor of safety needs to be applied to equation (6.6) to allow for the difference between a well-controlled anaerobic reactor and a septic tank (which is basically uncontrolled by comparison), so that t_d is given by:

$$t_d = 30\ (1.035)^{35-T} \quad (6.7)$$

The volume of fresh sludge is around 1 l per person per day. This is digested in t_d days when it passes to the sludge storage zone. So the average volume of digesting sludge present during the period t_d is 0.5 lcd. Thus, the volume of the sludge digestion zone (V_d, m^3) is given by:

$$V_d = 0.5 \times 10^{-3} P t_d \quad (6.8)$$

6.3.4 Sludge storage

The volume of the sludge storage zone depends on the rate of accumulation of digested sludge (r, m^3 per person per year) and the interval between successive desludging operations (n, years—see section 6.4). Measurements of sludge depth in septic tanks cannot distinguish between digesting sludge and digested sludge, so reported rates of sludge accumulation depend on the period of observation, as shown in Figure 6.3, from which the following design values for r are obtained:

For $n < 5$: $r = 0.06\ \mathrm{m^3/person\ year}$

and $n > 5$: $r = 0.04\ \mathrm{m^3/person\ year}$

The sludge storage volume (V_{sl}, $\mathrm{m^3}$) is given by:

$$V_{sl} = rPn \tag{6.9}$$

6.3.5 Overall design capacity

The overall design capacity of the septic tank (V, $\mathrm{m^3}$) is the sum of the volumes required for scum storage, sedimentation, digestion and sludge storage:

$$V = V_{sc} + V_h + V_d + V_{sl} \tag{6.10}$$

Since V_{sc} is around $0.4V_{sl}$, this becomes:

$$V = V_h + V_d + 1.4V_{sl} \tag{6.11}$$

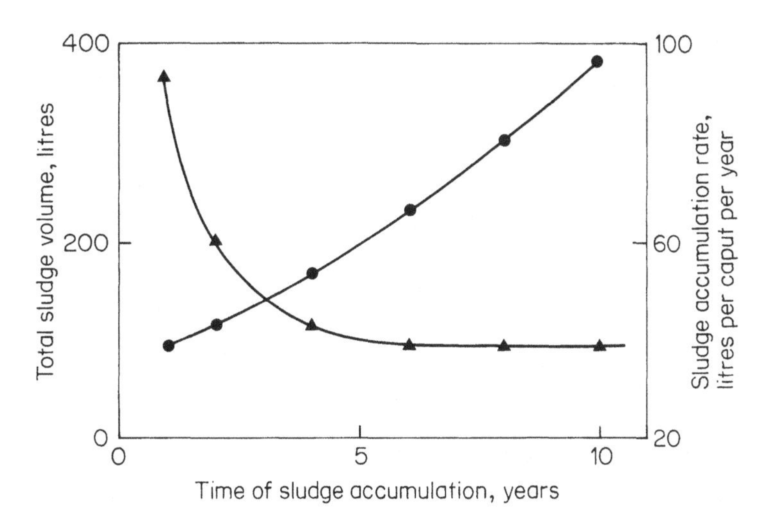

Figure 6.3 Variation in sludge volume (●) and sludge accumulation rate (▲) with time.

6.3.6 Two-compartment septic tanks

When the septic tank effluent is to be disposed of on-site (section 6.5), it is best to have a two-compartment tank (Figure 6.4); this minimizes the suspended solids content of the tank effluent, as any solids resuspended from the sludge layer in the first compartment by peak flows entering it, are able to settle again in the more quiescent second compartment. The overall effective tank volume (V, equation (6.11)) is normally divided into two-thirds for the first compartment and one-third for the second.

6.3.7 Communal septic tanks

Small housing blocks may advantageously be served by a single, communal septic tank; per household costs are considerably reduced (Figure 6.5). Each house is connected to the tank by in-block simplified sewers (see Chapter 9).

6.3.8 Design details

Geometry

The preferred shape of septic tanks is rectangular with a length-to-breadth ratio of 2–3 to 1, in order to reduce short-

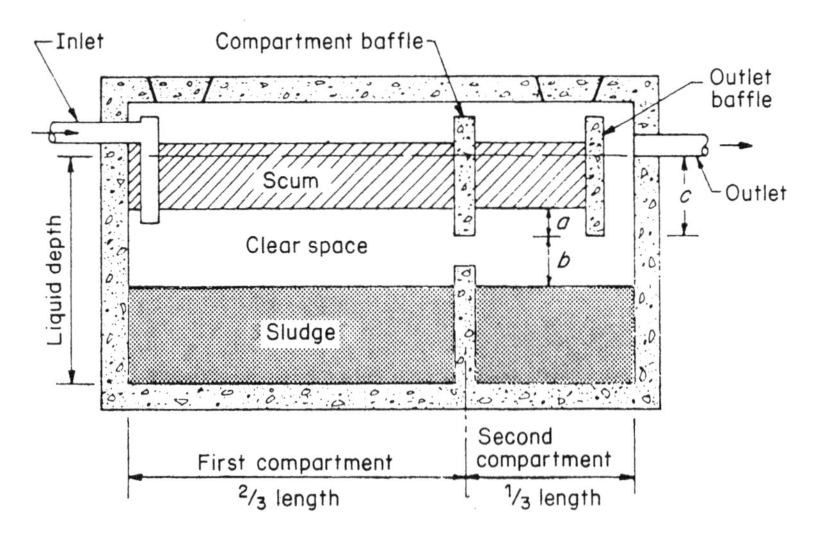

Figure 6.4 Two-compartment septic tank for use when the effluent is disposed of on-site.

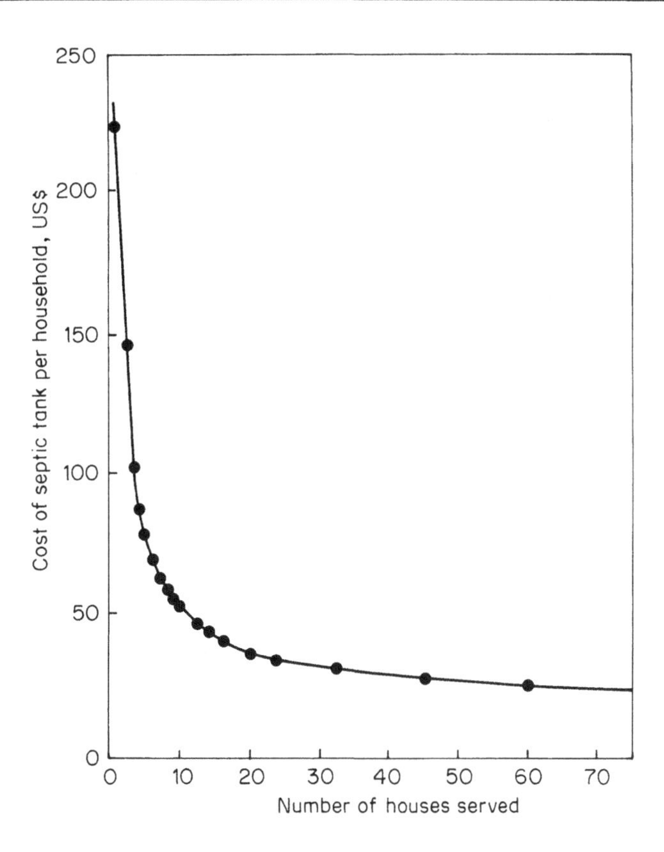

Figure 6.5 Variation of septic tank costs per household with the number of households served (data from Natal, northeast Brazil, 1983).

circuiting of the raw wastewater across the tank, and so improve suspended solids removal.

For equal volumes, shallow tanks are preferred to deep tanks, since they provide more hydraulic surge storage capacity and greater reduction in outflow velocity, so improving solids retention. Also, the depth of excavation is reduced, thus lowering costs and lessening problems with groundwater infiltration. The liquid depth of the tank should be at least 0.9 m, but not more than 2 m.

Inlets and outlets

The inlet to the tank should be the same size as the incoming house connection, usually 100 mm diameter. The tank outlet is also 100 mm diameter.

Baffles should be provided at both the inlet and outlet. Inlet baffles are designed to dissipate the energy of the influent and deflect it downwards into the tank. This prevents short-circuiting of the liquid across the top of the tank to the outlet. The outlet baffle is designed to retain the scum layer within the tank. Sanitary tees are normally used as baffles; they should extend 150 mm above the liquid level, so that they are above the scum layer, and down to around 30–40 percent of the liquid depth. The outlet invert should be at a sufficient level below that of the inlet to provide some surge storage and prevent stranding of solids in the building sewer during the momentary rises in the liquid level when wastewater enters the tank. A drop of 75 mm is recommended. A minimum freeboard or space above the liquid level of 300 mm should be provided for scum storage and ventilation.

Access

Manholes (300–600 mm) must be provided in the septic tank for removal of the accumulated sludge and scum. An inspection port above the inlet and outlet must also be provided; this allows cleaning of the baffle should it become blocked.

Structural design

Structural considerations should include soil loading and hydrostatic loading. Soil loadings must be considered for the tank walls, floor and cover, and in areas of high groundwater table hydrostatic loads on the empty tanks must also be included, to avoid it being moved vertically upwards during and immediately after desludging. Septic tanks are not usually located in areas subject to traffic loadings; but, if they have to be, the tank cover must be designed to resist collapse.

6.3.9 Upflow filter

The quality of septic tank effluent can be considerably improved by passing it through an upflow filter, which can be incorporated into the septic tank structure (Figure 6.6). The effluent enters at the base of the filter, flows upwards through a layer of coarse aggregate about 0.5 m deep and is discharged over a weir at the top. Anaerobic bacteria grow on the surface of the aggregate and oxidize the effluent as it passes through. The head loss is low, about 30–150 mm during normal operation. Field studies in India have shown that these filters can

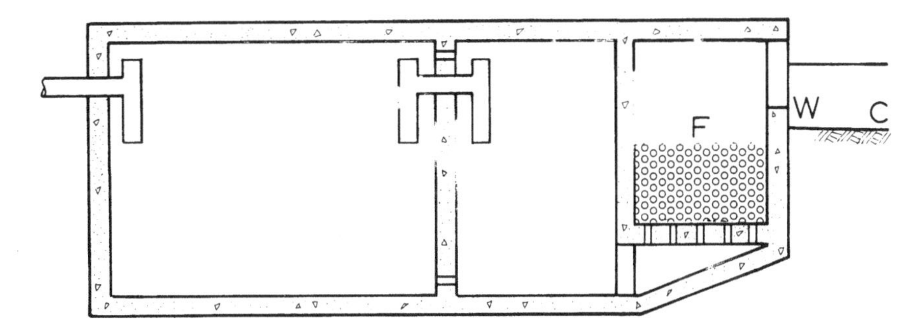

Figure 6.6 Septic tank with upflow filter. F, filter; W, effluent weir; C, effluent channel (or pipe to drainfield).

effect a 70 percent reduction in BOD and change a malodorous, highly turbid, grey-to-yellow influent to an odourless, clear, light yellow effluent. A filter capacity of about 0.05 m^3 per person is adequate and a satisfactory specification for the aggregate is:

Top 100 mm: 3–6 mm

Bottom 400 mm: 12–18 mm

An upflow filter may be expected to operate satisfactorily without maintenance for 18–24 months, when it becomes necessary to drain the filter and wash it with one or two flushes of clean water.

6.4 TANK DESLUDGING

The septic tank should be desludged (see Chapter 7) after n years where n is the chosen interval between successive desludging operations (equation (6.9)). The problem in practice is that the householder is most likely to forget to have the tank desludged, especially when $n > 2$—a survey done some years ago in California indicated that, in practice, n was over 70 years, i.e. most householders never had their tanks desludged! The result is that the tank fills with solids, and solids pass out into the drainfield and block it; effluent then works its way to the surface, and only then does the householder realise that something is wrong. To avoid all this happening, it is best to keep n small, 1 or 2 for example, but probably not greater than 3.

6.4.1 White towel test

A good way to measure sludge depth, and also the thickness of the scum layer, is to perform this simple test. White towelling material is wrapped around the lower half of a timber pole, which is then lowered right to the bottom of the septic tank via one of its desludging ports. The pole is then slowly and carefully withdrawn; some of the sludge particles will have become trapped in the towelling material, so making measurement of the sludge depth easy. The same pole can be used to measure the scum thickness if a large (100–150 mm) nail is hammered into it somewhere in its top half. The pole is lowered into the tank through the scum (this is best done when measuring the sludge depth), rotated and raised *slowly*: when the nail reaches the underside of the top of the scum layer, resistance can be felt; a mark is then made on the pole corresponding to the top of the scum layer, so that, once the

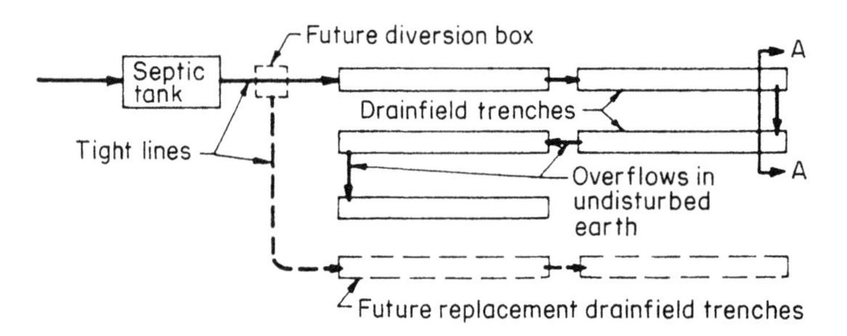

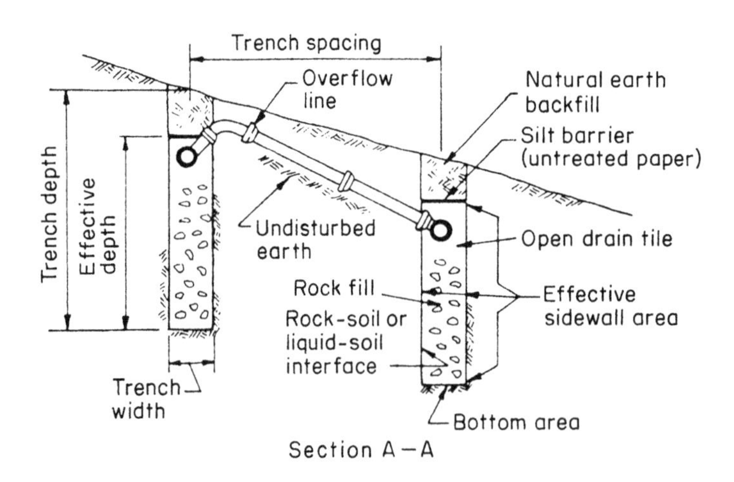

Figure 6.7 Drainfield trenches for septic tank effluent disposal.

pole is removed, the scum thickness is given by the distance between the nail and the mark.

6.5 ON-SITE EFFLUENT DISPOSAL

Septic tank effluent can be disposed of on-site in a drainfield, which comprises a series of drainage trenches (Figure 6.7). Each trench consists of open-joint agricultural drainage tiles of 100 mm diameter laid on a 1 m depth of rock fill (20–50 mm grading). The effluent infiltrates into the soil surrounding the trench, which eventually becomes clogged with sewage solids (provision must therefore be made to set aside land for use as a future replacement drainfield; the two drainfields can then be used alternately). The long-term infiltration rate is given in Table 4.1. If the soil is essentially impermeable, then the septic tank should be connected to a settled sewerage system (Chapter 8).

6.6 FURTHER READING

L. W. Carter and R. C Knox, *Septic Tank System Effects on Ground Water Quality*. Lewis Publishers, Chelsea MI (1985).

Construção e Instalação de Fossas Sépticas e Disposição dos Efluentes Finais. Brazilian Standard No. NBR 7229. Associação Brasileira de Normas Técnicas, Rio de Janeiro (1982).

J. M. de Azevedo Netto, "Tanques sépticos: conhecimentos atuais". *Engenharia Sanitária* (Rio de Janeiro) **24** (2), 222–229 (1985).

Design Manual: Onsite Wastewater Treatment and Disposal Systems. Report No. EPA 625/1-80-012. Environment Protection Agency, Cincinnati OH (1980).

R. J. L. C. Drews, *A Guide to the Use of Septic Tank Systems in South Africa*. CSIR Technical Guide No. K86. National Institute for Water Research, Pretoria (1986).

R. Laak, *Wastewater Engineering Design for Unsewered Areas*, 2nd edition. Technomic Publishing Co., Lancaster OH (1986).

D. D. Mara and G. S. Sinnatamby, "Rational design of septic tanks in warm climates". *The Public Health Engineer* **14** (4), 49–55 (1986).

R. J. Perkins, *Onsite Wastewater Disposal*. Lewis Publishers, Chelsea MI (1989).

G. Sagar, "A dwarf septic tank developed in India". *Waterlines* **2** (1), 22–23 (1983).

J. H. T. Winneberger, *Septic Tank Systems: A Consultant's Toolkit—Vol. 1: Subsurface Disposal of Septic Tank Effluencts; Vol. 2: The Septic Tank*. Butterworth Publishers, Stoneham MA (1984).

6.7 DESIGN EXAMPLE

Design a septic tank and drainfield to serve a family of six whose sewage production is 100 lcd. The design temperature is 25° C, and the soil is a sandy loam.

6.7.1 Tank design

1. *Sedimentation*. Use Equations (6.1) and (6.2):

$$\begin{aligned} t_h &= 1.5 - 0.3\log(Pq) \\ &= 1.5 - 0.3\log(6 \times 100) = 0.67 \text{ day} \\ V_h &= 10^{-3}\, Pqt_h \\ &= 10^{-3} \times 6 \times 100 \times 0.67 = 0.4 \text{ m}^3 \end{aligned}$$

2. *Sludge digestion*. Use equations (6.7) and (6.8):

$$\begin{aligned} t_d &= 30\,(1.035)^{35-T} \\ &= 30\,(1.035)^{35-25} = 42.3 \text{ days} \\ V_d &= 0.5 \times 10^{-3}\, Pt_d \\ &= 0.5 \times 10^{-3} \times 6 \times 42.3 = 0.13 \text{ m}^3 \end{aligned}$$

3. *Digested sludge storage*. Use equation (6.9), choosing n = 2 years and r = 0.06 m^3 per person per year:

$$\begin{aligned} V_{sl} &= rPn \\ &= 0.06 \times 6 \times 2 = 0.72 \text{ m}^3 \end{aligned}$$

4. *Overall effective tank volume*. Use equation (6.11):

$$\begin{aligned} V &= V_h + V_d + 1.4\, V_{sl} \\ &= 0.4 + 0.13 + (1.4 \times 0.73) = 1.5 \text{ m}^3 \end{aligned}$$

Use a two-compartment tank, with the first compartment 1 m^3 in volume and the second 0.5 m^3.

6.7.2 Drainfield design

From Table 4.1 the long-term infiltration rate for a sandy loam is 30 l/m^2 per day. The trench sidewall infiltration area is given by:

$$(\text{effluent flow, l/day})/(\text{infiltration rate, l/m}^2 \text{ per day}) = (100 \times 6)/30 = 20 \text{ m}^2$$

Take the effective trench depth as 0.7 m. The total trench length (remembering that the trench has two sides) is given by:

$$= \tfrac{1}{2}(20/0.7) = 15 \text{ m}$$

Use three trenches in series, each 5 m in length.

Note that, in calculating the trench length, only the sidewall area is considered. This is because the bottom of the trench becomes rapidly clogged, leaving the sidewalls as the only effective infiltration surface.

7

Emptying

7.1 MANUAL OR MECHANICAL EMPTYING?

When VIP latrine pits and PF toilets are full, to within 0.5 m or so of the coverslab, they require emptying. Septic tanks require regular desludging, at intervals of one to five years (section 6.4), as do the solids interceptor tanks used in settled sewerage (see Chapter 8).

If VIP latrines and PF toilets are of the alternating twin-pit variety (sub section 3.2.2), then the pits can be emptied manually, provided that they are dry pits (i.e. above the water table). There is no health risk to the workers as the contents of the pit are at least two years old and only a very few *Ascaris* ova will be viable; all other excreted pathogens will be dead. However, if the pits are wet, or if they are single pits, they must be emptied mechanically by vacuum tankers as their contents are not pathogen-free (there will be fresh excreta containing viable pathogens at the top of the pit). Septic tanks and solids interceptor tanks are also desludged by vacuum tankers.

7.2 VACUUM TANKERS

Vacuum tankers used to desludge septic tanks are widely available; they normally incorporate a standard sliding-vane vacuum pump, which is suitable for removing the light watery sludges from septic tanks. They may also be suitable for desludging wet pits if there are no bulky anal cleansing materials present in the sludge. However, in high-density urban areas access may be difficult.

More powerful vacuum tankers are needed to empty dry pits. High-performance vacuum tankers, with capacities of 9–18 m^3, have now been developed specifically for this purpose: for example, BREVAC tankers (Figure 7.1; see section

Figure 7.1 A BREVAC tanker used to desludge dry latrine pits

7.6). They incorporate a powerful liquid-ring vacuum pump capable of generating a vacuum of 0.85 bar (650 mm Hg); this, when combined with pneumatic conveying (Figure 7.2), means that pit sludges can easily be moved under suction up to 100 m horizontally, i.e. the tanker can be up to 100 m away from the pit being desludged, so access is not commonly a problem. If it is, then smaller vacuum tankers are available (for example, the Land Rover-mounted BREVAC-LA (Figure 7.3) and the Micravac tanker, both of which have a capacity of 1 m^3; see section 7.6).

A manually operated vacuum mini-tanker is also available, called the MAPET (*ma*nual *p*it latrine *e*mptying *t*echnology) system (Figure 7.4). It has the advantage that it can be manufactured locally at low cost (around US$ 3000). It comprises a 0.2 m^3 tank (made from 3 mm mild steel sheet) mounted on a handcart incorporating tricycle wheels (overall width: 800 mm), and a vacuum pump made from 150 mm PVC sewer pipe with a leather piston and an 800 mm diameter flywheel. The MAPET system was developed in Dar es Salaam, where it is operated by the informal sector in close cooperation with the municipal sewerage and sanitation department.

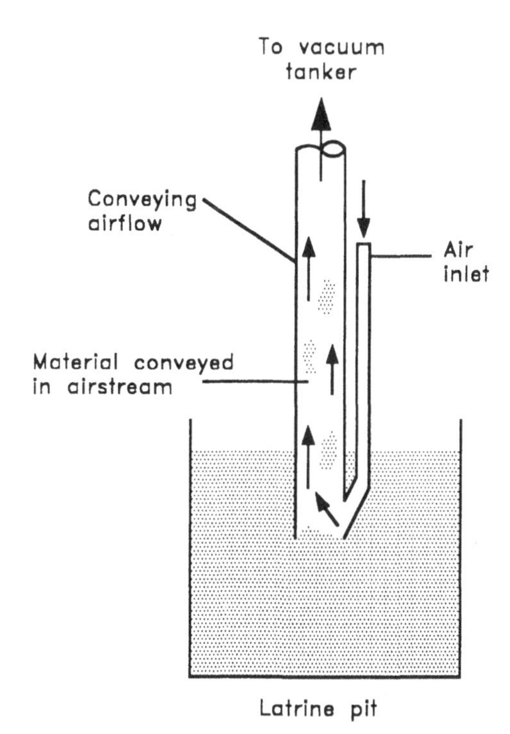

Figure 7.2 The principle of pneumatic conveying: the pit solids are conveyed in a stream of air

Figure 7.3 The BREVAC-LA tanker used in areas of limited vehicular access

7.3 SLUDGE DISPOSAL

Sludges from alternating twin pits are essentially pathogen-free, so disposal could be either on-site if the householder so

Figure 7.4 The MAPET system for pit desludging

wishes (to add to a vegetable garden, for example), or off-site (to farmland, treebelt or forest, sanitary landfill).

Sludges from single pits, septic tanks and solids interceptor tanks are not microbiologically safe, so disposal without prior treatment is only possible to sanitary landfills (and the sludge must be quickly and completely covered with garbage, not least to protect the health of any garbage-pickers or scavengers), or treebelts and forests to which the public does not have access. Treatment is required before application to agricultural land or aquacultural ponds (see Chapter 10). Composting is also feasible.

7.4 INSTITUTIONAL ASPECTS

Public sector organizations (such as municipalities, city councils) do not enjoy a high reputation for vacuum tanker maintenance—it is all too common to find possibly only one or two out of a dozen tankers operational. Yet the skills required for vehicle maintenance are not lacking in developing countries, as witnessed by the operational success of private sector haulage companies, for example. The public sector organization responsible for emptying pits and desludging septic and solids interceptor tanks has to decide whether to do the job properly itself (and this means organizing and undertaking efficient emptying schedules; being able to pay for vehicle fuel; training its staff in vehicle maintenance; ensuring an adequate stock of lubricants and spare parts), or whether to employ a private sector organization to do it under contract. In either case, access to foreign exchange (to purchase vacuum tankers and spare parts) is required.

7.5 EMPTYING COSTS

Vacuum tankers are not cheap: for example, a 5 m^3 capacity BREVAC tanker costs around US$ 120 000 ex works. However, the cost per pit emptied is not high, and the equivalent monthly charge to householders for pit emptying is really quite small, as the following calculation shows.

Suppose that the delivered cost of the tanker is US$ 125 000; its annual operation and maintenance costs are US$ 10 000; and it can empty 10 pits a day for 200 days a year. Suppose, further, that the pits have an emptying frequency of two years, and the tanker costs are written off over 10 years. The annual costs of providing the tanker service are as follows:

- tanker costs ($125 000 amortized at, say, 15 percent over 10 years—see Chapter 12): $25 000
- operation and maintenance: $10 000

i.e. an annual total of $35 000. During the year the tanker empties 2000 pits, so the cost per pit emptied is $17.50. The pit has to be emptied every two years, so the monthly cost per household is $0.73. These costs are only order-of-magnitude estimates but fairly reasonable nonetheless: actual household emptying costs per month in Rio de Janeiro in 1986 were $0.40–0.50. Pit-emptying costs are thus affordable under most circumstances, provided that the up-front (foreign exchange) costs can be met (these would normally be part of a loan package).

7.6 FURTHER READING AND INFORMATION

A. Boesch and R. Schertenleib, "Empting on-site excreta disposal systems: field tests with mechanized equipment in Gabarone (Botswana)". *IRCWD News No. 21/22*, pp. 1–10. International Reference Centre for Wastes Disposal, Duebendorf (1985).
R. F. Carroll, *A Mechanised Method of Emptying Sanitation Chambers*. Information Paper No. 1/84. Building Research Establishment, Watford (1984).
R. F Carroll, *Mechanised Emptying of Pit Latrines*. Overseas Building Note No. 193. Building Research Establishment, Watford (1989).
M. S. Muller and J. Rijnsburger, "MAPET: An appropriate latrine-emptying technology". *Waterlines* **13** (1), 24–26 (1994).
E. I. Stentiford, *Low-cost Composting of Solid Wastes*. TPHE Research Monograph No.4. University of Leeds (Department of Civil Engineering), Leeds (1995).
M. Strauss and U. Heinss, "SOS—Management of sludges from on-site sanitation: faecal sludge treatment". *SANDEC News No.1*. pp. 2–13. Water and Sanitation in Developing Countries, Duebendorf (1995).

Information on the range of BREVAC vacuum tankers is available from the manufacturers:

Brain Industries Ltd
Woodcock Wells
Kilgetty, Dyfed SA68 0UJ
United Kingdom

Fax: +44 1834 811878

and on the Micravac vacuum tankers from:

MCA Vehicles Ltd
Kiltipper Road
Tallaght
Dublin 24
Ireland

Fax: +353 1 519604

Further details of the MAPET system are available from:

WASTE Consultants
Crabethstraat 38F
2801 AN Gouda
The Netherlands

Fax: +31 1820 84885

8

Settled Sewerage

8.1 DESCRIPTION

Settled sewerage (Figure 8.1) is a sewerage system that is designed to receive only the liquid fraction of household wastewater; settleable solids, grease and scum are removed in a solids interceptor tank (basically a single-compartment septic tank), which is installed upstream of each connection to the sewer (or groups of connections, the interceptor tank being shared by adjacent houses).

Because the sewers receive only settled sewage, they are designed very differently from conventional sewers. The most

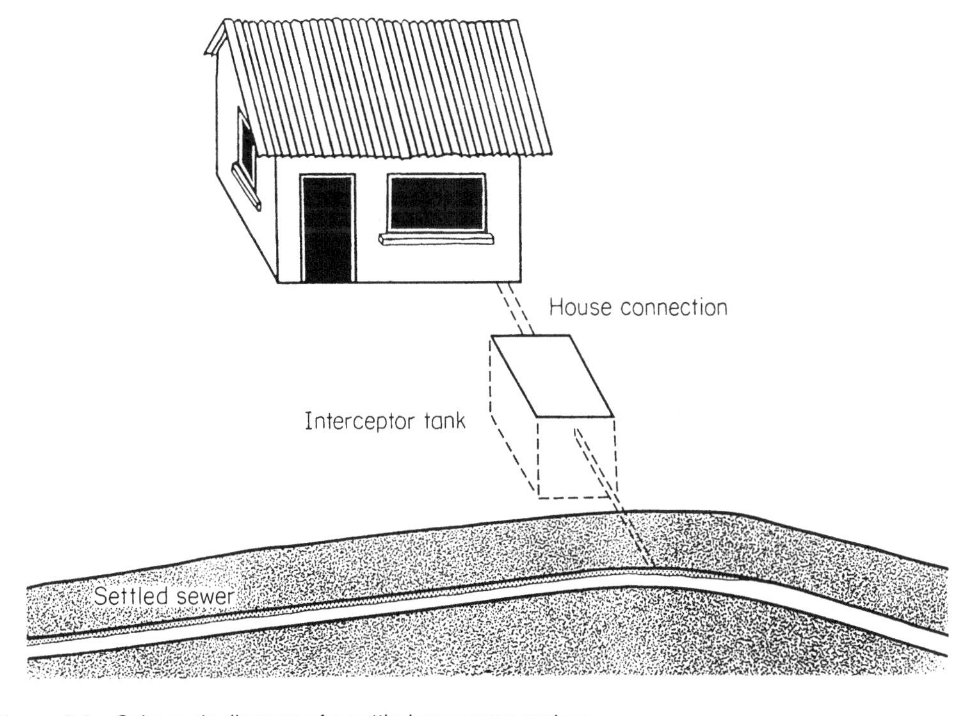

Figure 8.1 Schematic diagram of a settled sewerage system

obvious differences are that they do not have to be designed for self-cleansing velocities (i.e. velocities to ensure conveyance of solids), and that the flow in the sewers can vary from normal gravity open channel flow to full-bore pressure flow and then back to open channel flow (see subsection 8.3.3). By comparison with conventional sewerage settled sewerage costs are quite low and this is mainly due to its shallow excavation depths and the use of small diameter pipework (commonly 75–100 mm PVC) and simple inspection units in place of large manholes.

Settled sewerage is also called small-bore sewerage or small diameter gravity sewerage, but these terms could also describe simplified sewerage (Chapter 9). Settled sewerage is a more precise description of the system in which sewage settled in the interceptor tank is conveyed off-site to a treatment works (Chapter 10). Solids-free sewerage could be an alternative term (in Portuguese the system is called *redes de esgotos decantados*, in French *réseaux d'eaux usées decantées*, and in Spanish *alcantarillado sin arrastre de sólidos*).

8.2 APPROPRIATENESS

Settled sewerage is most appropriate for areas that already have septic tanks but where the soil can no longer accept all the septic tank effluent. So it is often a lower-cost solution in middle- to upper-income areas where it is used as an alternative to more costly conventional sewerage. Saving money in this way means (or should mean) that there is more available to serve low-income areas. Settled sewerage can also be used as a means to upgrade pour-flush toilets (Chapter 13).

However, if the reason why the soil can no longer accept septic tank effluent is simply because of a high in-house water consumption (≫ 100 litres per caput per day, lcd), with a correspondingly high wastewater generation, then serious consideration should be given to in-house water conservation techniques, such as the installation of water-saving plumbing fixtures (for example, low-volume flush toilets, flow restrictors on taps and showerheads), in order to reduce the resulting wastewater flow such that the soil is again able to accept the septic tank effluent.

Settled sewerage is commonly used in Australia (where it is called common effluent drainage), and also in the United

States, Columbia, Nigeria and Zambia. Its increasing use in the United States for new housing developments is simply due to its low cost (around 50–60 percent of conventional sewerage) and the fact that, from the users' perspective, there is little noticeable difference between it and conventional sewerage.

8.3 DESIGN

8.3.1 Interceptor tank

The solids interceptor tank is designed in exactly the same way as a septic tank (Chapter 6). Single compartment tanks are sufficient (rather than two-compartment ones) as any suspended solids leaving the tank will not block the sewer. Costs can be reduced by a group of houses sharing a single tank (see Figure 6.5).

8.3.2 System layout

A settled sewerage system is a means by which settled wastewater from all households is conveyed to a selected outlet utilizing the energy resulting from the difference in elevation between its upstream and downstream ends. It must be deep enough in the ground to receive the flows from each house and it must have sufficient size and gradient to carry these flows. Therefore, hydraulic losses have to be within the limits of available energy. Where the differences in elevation are insufficient to permit gravity flow, energy must be added to the system by lift pumps. The number and location of lift stations is generally determined from comparisons of their costs of construction, operation and maintenance with the cost of construction and maintenance of deeper and/or larger diameter sewers. The consequences of mechanical or electrical breakdown of lift stations to the safety and convenience of the community must also be taken into account.

A system layout can be largely made from good 1-in-2000 topographical maps of the area to be served as these show elevations, existing roads, buildings, property boundaries and other relevant information. The layout begins by selecting the system outlet and defining service district and sub-district boundaries. Wherever possible, the district and sub-district boundaries are made to conform to natural drainage basins. Within these boundaries, the branch and main sewer routes are selected. Selection of sewer routes must consider the following:

- interceptor tank locations and elevations
- vertical and horizontal alignments
- rights-of-way and easements
- lift stations
- future development, and
- construction disruption and site restoration.

The location and outlet elevation of the interceptor tanks, together with the local topography, will generally establish the route and depth of the sewers. Existing rights-of-way and easements should be used, but if excavation costs can be reduced significantly by some other route, special easements may be necessary. Whilst it is preferable that every connection be served by gravity, the local terrain or cost of excavation may require that lift stations be used; groups of households can often be served by a single lift station. The costs of reinstating pavements, curbs and gutters and other structures disturbed during construction is an important consideration in locating routes; curvilinear alignment will allow some structures to be avoided (but joint deflections must not exceed those permitted by the pipe manufacturer). Also, if the properties to be served are on both sides of a roadway, it may be cheaper to lay a sewer on each side in order to avoid expensive road crossings.

8.3.3 Peak flow estimation

The wastewater flow entering the sewer is markedly attenuated in the interceptor tank. The extent of attenuation is a function of the tank's surface area and the length of time during which the wastewater is discharged to the tank. However, there are very few data on the magnitude of peak flows in settled sewers. In the United States, peak factors of 1.2–1.3

have been observed. A suitable design value is probably 1.5. Thus, the peak flow from each household (q_h, l/s) is given by:

$$q_h = 1.5\ kpw/86\,400 \tag{8.1}$$

where k = return factor (sewage flow/water consumption), typically 0.8–0.9
p = household size
q = water consumption, lcd

and 86 400 is the number of seconds in a day.

Thus, for $k = 0.85$, equation (8.1) can be rewritten as:

$$q_h = 1.5 \times 10^{-5} pw \tag{8.2}$$

The peak flow at the downstream end of a sewer section is simply the peak flow per household multiplied by the number of households discharging into that section (see the design example in section 8.7). The peak flow from a lift station (subsection 8.4.2) is equal to the pump discharge rate.

An estimate of groundwater infiltration and surface water inflow must be made. Ideally, of course, any such 'clear' water entry should be zero, but in practice some imperfectly sealed pipe joints must be expected; these are less common with PVC pipes, as the joints are easier to make and pipe lengths are longer than with vitrified clay pipes. Conservative estimates of clear water inflow are around 20 m^3/ha per day for vitrified clay pipes and 10 m^3/ha per day for PVC pipes.

8.3.4 Hydraulic design

The flow in settled sewers can be either open channel gravity flow or, in sections that are below the hydraulic gradient, full-bore pressure flow. In the hydraulic design of settled sewers, separate analysis has to be made for each sewer section in which the type of flow (open channel or pressure) does not vary and the hydraulic gradient is resonably uniform.

Manning's equation can be used:

$$v = (1/n) r^{2/3} i^{1/2} \tag{8.3}$$

where v = velocity of flow, m/s
n = pipe roughness coefficient

r = hydraulic radius (area of flow/wetted perimeter), m
i = pipe slope (strictly, the hydraulic gradient), m/m

Where the sewer is flowing just full, $r = R$ and the value of R is given by:

$$R = (\pi D^2/4)/\pi D$$

i.e.

$$R = D/4 \tag{8.4}$$

where D = sewer diameter, m.

Since velocity is flow/area, the flow in the sewer when flowing just full is given by equations (8.3) and (8.4) as:

$$Q/(\pi D^2/4) = (1/n)(D/4)^{2/3} i^{1/2} \tag{8.5}$$

The value of n is usually 0.013 (but see subsection 9.3.3), so equation (8.5) can be rearranged to give:

$$Q = 24 D^{8/3} i^{1/2} \tag{8.6}$$

Changing the units of Q from m^3/s to l/s and of D from metres to millimetres gives:

$$Q = 2.4 \times 10^{-4} D^{8/3} i^{1/2} \tag{8.7}$$

The value of Q given by equation (8.7) must be greater than the estimated peak flow in the section of sewer under consideration; if it is not, either the next larger pipe size is used, or the pipe gradient is increased by increasing the depth of excavation along the section. The design example given in section 8.7 shows how equation (8.7) is used.

Minimum diameter

In the United States a minimum pipe diameter of 50 mm has been used successfully in settled sewerage schemes. However, to provide greater ease of sewer cleansing (should the need for this arise), a minimum diameter of 75 mm is recommended.

Inflective gradient design

The profile of a settled sewer can closely follow the ground profile, with flow conditions changing as required from open channel flow to pressure flow and back to open channel flow. This is termed the inflective gradient design approach, and is possible with settled sewerage, since maintenance of strict sewer gradients is not necessary to ensure self-cleansing velocities as the sewer receives only settled sewage. Clearly, the design must ensure that an overall fall does exist across the system, and that the hydraulic gradient does not rise above the level of the outlet invert of any interceptor tank—otherwise sewage would flow from the sewer into the tank. High points where the flow changes from open channel flow to pressure flow, and points at the end of long flat sections, are *critical points* where the maximum sewer elevation has to be established. Between these critical points the sewer can be laid at any profile as long as the hydraulic gradient is below the outlet invert of all interceptor tanks, and no additional high points are created (see the design example in section 8.7).

8.4 CONSTRUCTION AND MAINTENANCE

8.4.1 Sewer appurtenances

Simple cleanout units (Figure 8.2) to provide access can be used in most locations instead of manholes (manholes are expensive and also act as points of entry for grit and infiltration water). Cleanouts should be provided at:

- all upstream ends of the sewer
- all high points
- major changes in sewer direction
- major sewer intersections
- intervals of 150–200 m in long and flat sections.

At some major sewer intersections manholes may be necessary, rather than cleanouts, but the number of manholes should be kept to an absolute minimum.

Ventilation is not required for the satisfactory operation of

Figure 8.2 A typical cleanout unit: note that not all sewer junctions need to be provided with a cleanout

settled sewers if the sewers are laid on a continuous negative gradient. However, in sections where the flow is pressure flow, air may accumulate at high points and so the cleanout located there should be ventilated to allow any air (but not sewage) to escape.

8.4.2 Lift stations

Lift stations may be required at individual (or group) household connections where the interceptor tank outlet invert is below the sewer (and where deeper excavation would be more expensive than providing a lift station). These individual lift stations (Figure 8.3) are simple in design with low-head, low-capacity corrosion-resistant pumps; as the settled sewage contains no solids, the pumps are clean water pumps, rather than the usual (and more expensive) solids-handling sewage pumps. The pumps are controlled by mercury float switches set for small pumping volumes (this prevents the sewers becoming surcharged). A high-level alarm switch, operated off a separate circuit, should be set at around 150–200 mm above the pump-on switch to warn the users of any pump malfunction by a visual and audible alarm signal.

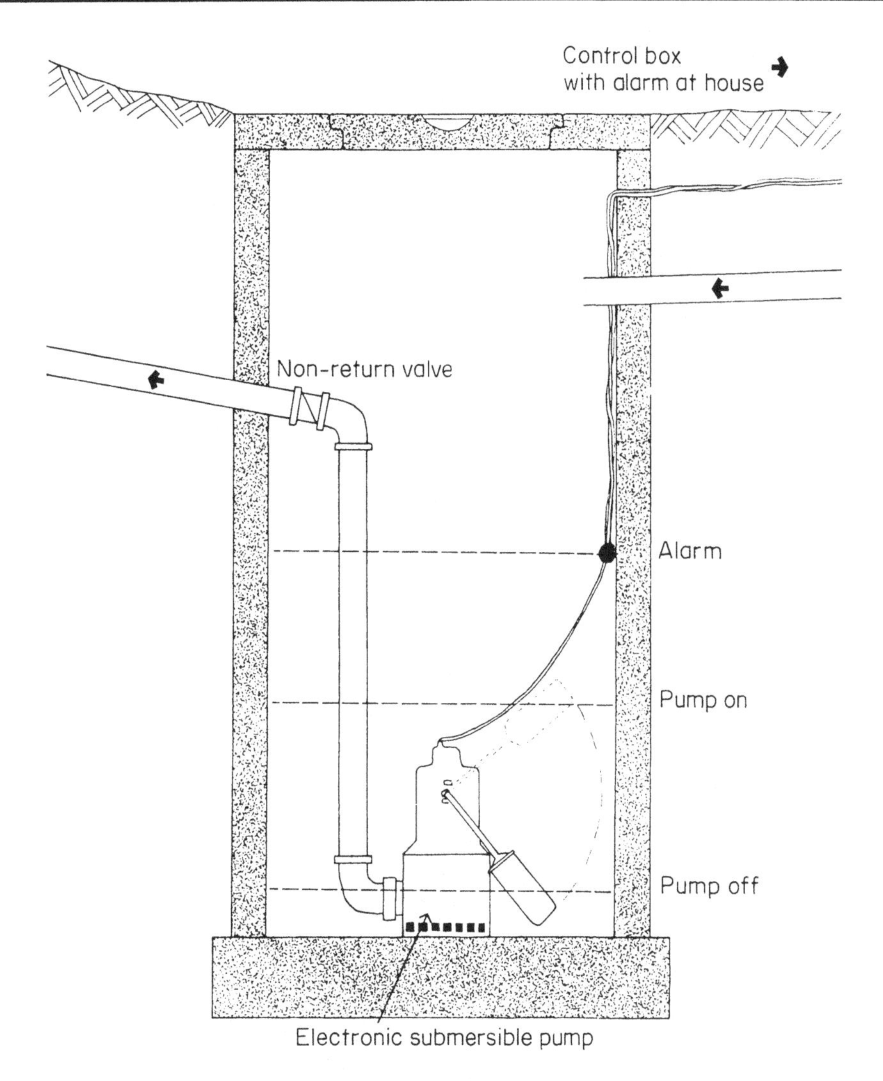

Figure 8.3 Lift station for use at a household connection

Lift stations serving a drainage basin or sub-basin (Figure 8.4) are conventional in design, except that large-capacity, solids-handling pumps are not required: again, less expensive corrosion-resistant clean water pumps are satisfactory. Because settled sewage soon becomes septic, corrosion and odour are both potentially problematic: any concrete should be of good quality made with a sulphate-resisting cement. To control odours, the settled sewage should discharge into the sump below low-water level via a drop inlet (basically an extended sanitary tee), and a fresh air vent to the sump should

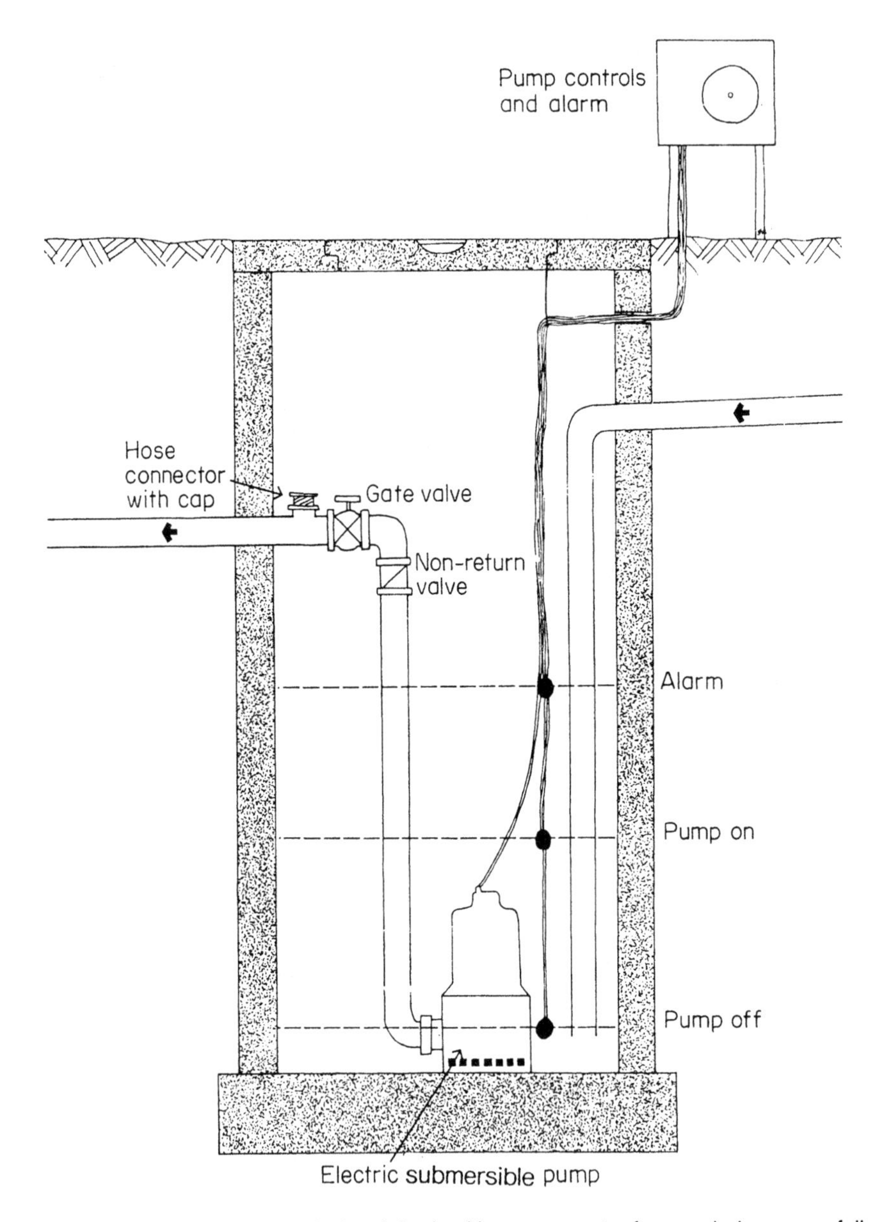

Figure 8.4 Main in-line lift station with drop inlet (and hose connector for use during power failure)

be provided. In the event of power failure, emergency storage is provided in the sump, and a truck-mounted, self-priming pump used to pump the sump contents directly into the pressure sewer by means of the hose connection shown in Figure 8.4.

8.4.3 Sewage treatment

A settled sewer network can join a conventional sewer if one is conveniently located nearby, or it can discharge into its own sewage treatment works (Chapter 10). If waste stabilization ponds are used, then anaerobic ponds are not required, as the interceptor tanks fulfil their function, and the settled sewage is treated in a series of facultative and maturation ponds. Care must be taken to avoid odour release from the septic settled sewage.

8.4.4 Operation and maintenance

The local sewerage authority has to be responsible for the operation and maintenance of the settled sewer network: this includes removal of any blockages (which are extremely rare), and maintaining the lift stations. It has to ensure that all connections to the system are *only* settled sewage connections from an interceptor tank, and it must be extremely vigilant that illegal connections without interceptor tanks do not occur (otherwise unsettled sewage would enter the network and block it).

Desludging of the interceptor tanks (Chapter 7) cannot be left to the householders: they simply cannot be relied upon to desludge on time, and settleable solids would enter the system. The responsibility for desludging *has* to be assumed by the sewerage authority, which can more easily arrange for all the interceptor tanks along a section to be desludged at the same time; costs are recovered by including them in the householders' monthly or quarterly bills. The sewerage authority is also. of course, responsible for the operation and maintenance of any sewage treatment works (Chapter 10).

8.5 FURTHER READING

South Australia Health Commission, *Common Effluent Drainage*. SAHC (Health Surveying Services), Adelaide (1982).

D. D. Mara, *The Conservation of Drinking Water Supplies: Techniques for Low-income Settlements*. United Nations Centre for Human Settlements, Nairobi (1989).

D. D. Mara (ed.), *Low-cost Sewerage*. John Wiley & Sons, Chichester (1996, forthcoming).

R. J. Otis and D. D. Mara, *The Design of Small Bore Sewer Systems*.

TAG Technical Note No. 14. The World Bank, Washington, DC (1985).
R. A. Reed, *Sustainable Sewerage: Guidelines for Community Schemes*. IT Publications, London (1995).
J. H. Rizo Pombo, "Alcantarillado de bajo costo: el sistema ASAS (alcantarillado sin arrastre de sólidos) en Cartagena". In *Tecnológias Urbanas Socialmente Apropriadas: Experencias Colombianas* (J. J. Guibbert (ed), volume 2, pp. 289–324. Documentos Tercer Mundo No. 47–48–49. ENDA-América Latina, Bogotá (1987).
L. J. Vincent, W. E. Algie and G. v. R. Marais. "A system of sanitation for low cost high density housing". In *Proceedings of the Symposium on Hygiene and Sanitation in Relation to Housing CCTA/WHO, Niamey 1961*, Publication No. 84, pp. 135–172. Commission for Technical Cooperation in Africa South of the Sahara, London (1963).

8.6 DESIGN EXAMPLE

A settled sewer is to be designed serving 10 compounds, with allowances for a future extension serving a further 10 compounds at the upstream end of the sewer, and for a future branch line serving 20 compounds, which discharges into the sewer just upstream of station 7 (see Figure 8.5). The average compounds size is 24 people and the average water consumption is 70 l/caput day.

8.6.1 Solution

The peak flow from each compound is given by equation (8.2):

$$\begin{aligned} q_{\mathrm{h}} &= 1.5 \times 10^{-5} pw \\ &= 1.5 \times 10^{-5} \times 24 \times 70 \\ &= 0.025\ \mathrm{l/s} \end{aligned}$$

Individual sewer sections are selected for hydraulic analysis on the basis of each section having relatively uniform gradients or flows. Here, nine sections are chosen, as shown in Figure 8.5. The hydraulic calculations are presented in Table 8.1 and described, column by column, below:

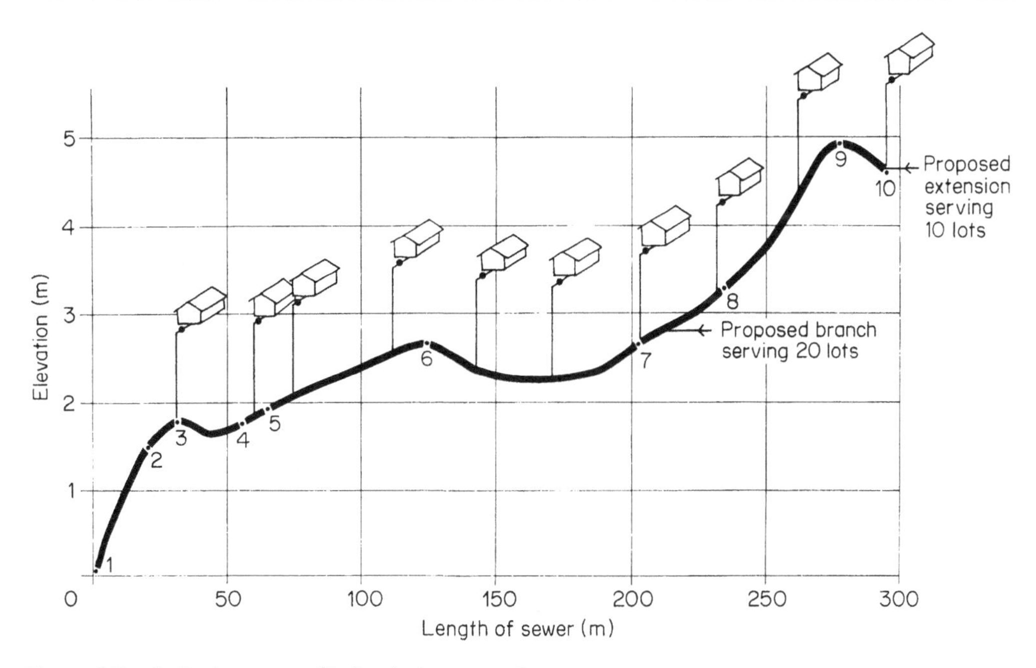

Figure 8.5 Settled sewer profile for design example

Column (1): station number

Numbers are given to the points at which each sewer section commences, starting from the downstream end of the sewer (see also Figure 8.5).

Column (2); station elevation

The elevation (m) of each station above datum (the datum in this example is taken as the elevation of Station 1).

Column (3): distance

The horizontal distance (m) of each station from Station 1.

Column (4): elevation difference over section

The difference (m) between the elevations (given in column (2)) of adjacent stations.

Table 8.1 Hydraulic calculations for design example

(1)	(2)	(3)	(4)	(5)	(6)	(7)	(8)	(9)	(10)	(11)
Station number	Station elevation (m)	Distance from station 1 (m)	Difference in elevation over section (m)	Length of section (m)	Average slope of section (m/m)	Number of compounds served	Design flow (l/s)	Pipe diameter (m/m)	Flow at full pipe (l/s)	Comments
1	0.00	0								
			1.65	21	0.079	40	1.00	50	2.29	
2	1.65	21								
			0.12	9	0.013	40	1.00	50	0.92	50 mm diameter is too small; use 75 mm: flow at full pipe = 2.74 l/s
3	1.77	30								
			0.00	24	0.000	39	0.98	50	–	
4	1.77	54			0.015[a]				1.00	Hydraulic gradient rises 0.36 m above station 4
			0.18	9	0.020	39	0.98	50	1.15	
5	1.95	63								
			0.67	64	0.010	38	0.95	50	0.81	50 mm diameter is too small; use 75 mm: flow at full pipe = 2.40 l/s
6	2.62	127								
			0.00	74	0.000	36	0.90	50	–	
7	2.62	201			0.012[a]				0.90	Hydraulic gradient rises 0.89 m above station 7
			0.61	34	0.018	34	0.85	50	1.09	
8	3.23	235								
			1.71	43	0.040	13	0.33	50	1.63	
9	4.94	278								
			−0.28	17	−0.017	11	0.28	50	–	
10	4.66	295			0.0012[a]				0.28	Hydraulic gradient rises 0.30 m above station 10[b]

[a] Slope of hydraulic gradient (calculated from equation (8.8)). Flow at full pipe calculated from equation (8.7) with this gradient.
[b] $[(0.0012 \times 17) + 0.28]$, i.e. $[(\text{col.}(6) \times \text{col.}(5)) + |\text{col.}(4)|] = 0.30$ m.

Column (5): section length

The difference (m) between the station distances (given in column (3)) of adjacent stations.

Column (6): average slope of section

Column (4) divided by column (5) gives the average section slope (m/m).

Column (7): number of connections served

The number of compounds connected upstream of the downstream station of the section.

Column (8): design flow

Column (7) multiplied by 0.025 (the peak flow in l/s per compound) gives the peak flow (l/s) in the section.

Column (9): sewer diameter

The diameter (mm) selected by the designer for each section (initial choices may be inadequate and the pipe size may have to be increased).

Column (10): flow at full pipe

Use equation (8.7) with i and D as given in columns (6) and (9), respectively. For example, for the first section:

$$\begin{aligned} Q &= 2.4 \times 10^{-4} D^{8/3} i^{1/2} \\ &= 2.4 \times 10^{-4} (50)^{8/3} (0.079)^{1/2} \\ &= 2.29 \text{ l/s} \end{aligned}$$

Critical sections

Critical sections in settled sewer design are those laid flat (i.e. zero gradient) and those subject to pressure flow. In this example, no sections are laid flat, but three are subject to pressure flow: the sections between stations 3 and 4, 6 and 7, and 9 and 10. These sections have to be analysed carefully to ensure there is no backflow from the sewer to an interceptor tank. To check this, proceed as follows:

- Use equation (8.7) to calculate the hydraulic gradient for the pipe flowing full.

For example, for the section between stations 6 and 7, with $Q = 0.90$ l/s and $D = 50$ mm:

$$Q = 2.4 \times 10^{-4} D^{8/3} i^{1/2}$$

i.e.

$$i = 1.736 \times 10^{7} Q^{2} D^{-16/3} \qquad (8.8)$$

$$= 1.736 \times 10^{7} (0.9)^{2} (50)^{-16/3}$$

$$= 0.012 \text{ m/m}$$

- Calculate the maximum elevation to which the hydraulic gradient rises.

In this section, the hydraulic gradient rises (0.012×74) m, where 74 is the section length in m, i.e. 0.89 m above the upstream end. The invert of the outlet of any interceptor tank discharging into the sewer along this section must be above the hydraulic gradient to avoid backflow into the tank during periods of peak flow.

Note that in two sections, those between stations 2 and 3 and stations 5 and 6, 50 mm diameter pipe is too small and 75 mm diameter pipe is used. Note also that in both these cases the 75 mm diameter pipe discharges into a 50 mm diameter pipe—this is one of the differences between settled sewerage and conventional sewerage (in which pipe diameters of downstream sections are always the same as, or larger than, those of upstream sections), and is possible only because all solids have been removed in the interceptor tanks.

The above example was done with an initial choice of pipe diameter of 50 mm. In practice, 75, mm diameter pipe would be used for the whole length of sewer.

9

Simplified Sewerage

9.1 DESCRIPTION

Simplified sewerage is a sewerage system that is designed to receive all household wastewater without settlement in solids interceptor tanks. It is essentially similar to conventional sewerage, but without any of the latter's conservative features (which have accrued in codes of design practice over the last 100 years or so). Small diameter sewers laid at shallow gradients are used to convey the sewage; these sewers are often laid inside housing blocks (Figure 9.1), when the system is known as *condominial sewerage*; or they may be laid outside the

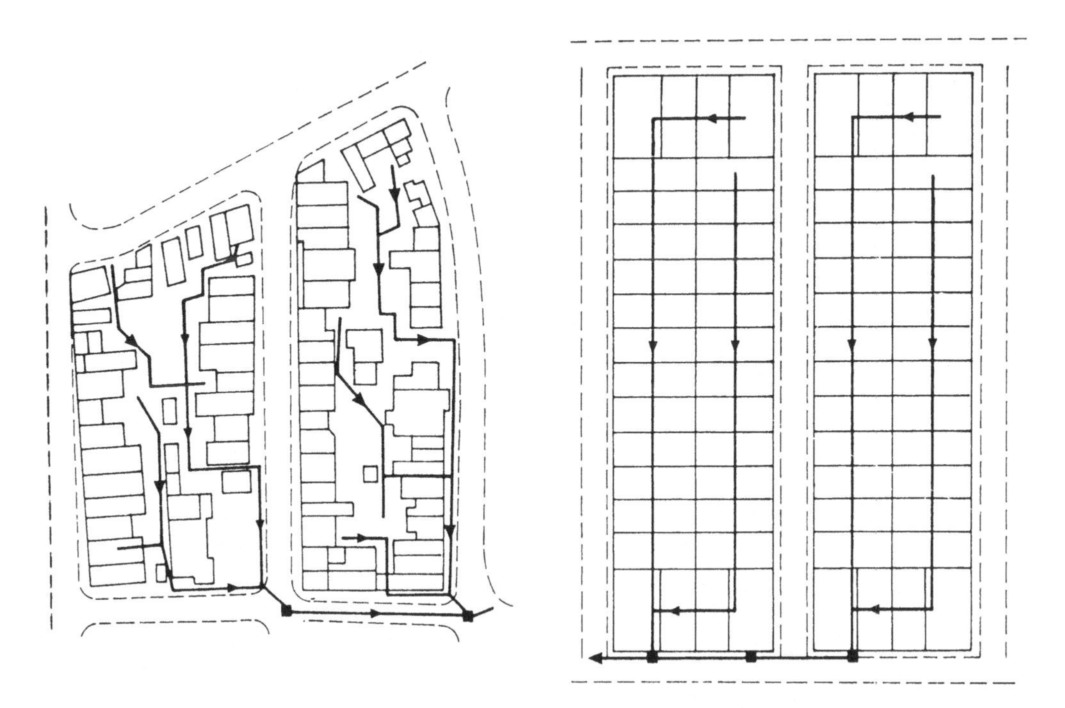

Figure 9.1 Sample layouts of simplified (condominial) sewerage for unplanned and planned housing areas

block, usually under both pavements (sidewalks) rather than in the middle of the road, as is the case with conventional sewerage.

Costs are low (section 9.5), and can even be lower than on-site sanitation (Figure 9.2). The low costs of simplified (especially condominial) sewerage are due, as in the case of settled sewerage (Chapter 8), to shallow excavation depths, small diameter pipework and simple inspection units in place of large manholes.

Simplified sewerage has also been called shallow sewerage, but this is not a good description as the sewers in settled sewerage (Chapter 8) are also laid at shallow depths (and simplified sewerage uses small diameter sewers, which is why small-bore sewerage is no longer a good description of settled sewerage). (In Portuguese, simplified sewerage is called *redes*

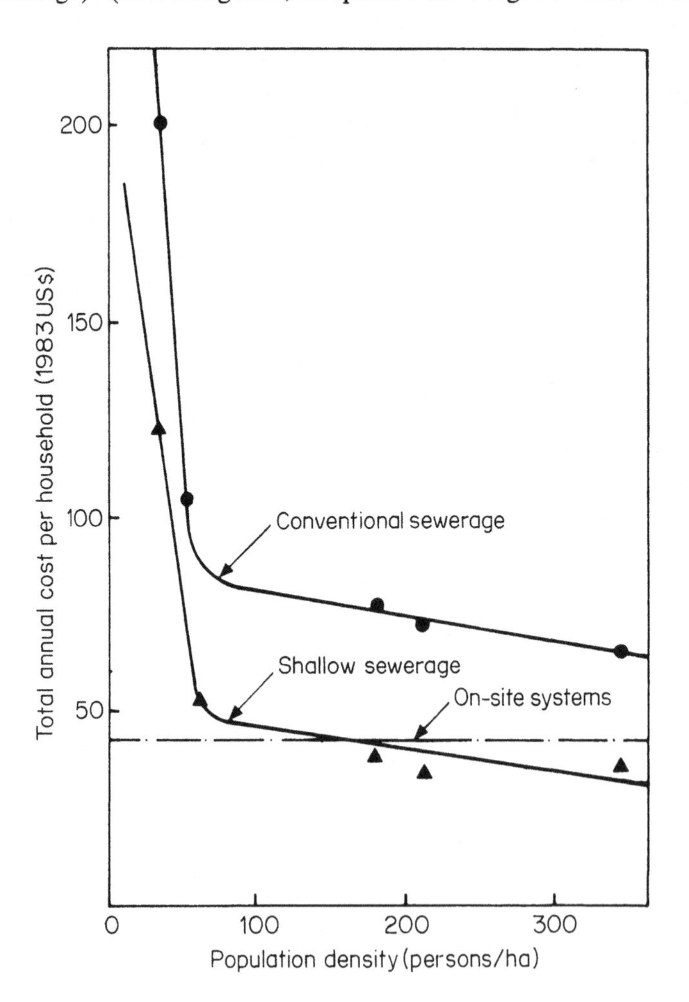

Figure 9.2 Costs of conventional, shallow (i.e. simplified, and, in fact, condominial) sewerage and on-site sanitation in Natal, northeast Brazil as a function of population density. Simplified sewerage became cheaper than on-site sanitation in this case at a population density greater than 160 persons per ha

de esgotos simplificadas, in French *réseaux d'eaux usées simplifiés*, and in Spanish *alcantarillado simplificado*.)

9.2 APPROPRIATENESS

Simplified sewerage is most appropriate in high-density, low-income housing areas where there is no space for on-site sanitation pits or for the solids interceptor tanks of settled sewerage. It was developed as condominial sewerage in the early 1980s by CAERN, the water and sewerage company of the state of Rio Grande do Norte in northeast Brazil, as an affordable solution to the up-to-then intractable problem of how to provide sanitation in high-density, low-income areas. It works well simply because it serves a high-density area: there should be a high initial rate of connection to the network (this is often well over 90 percent; with conventional sewerage it can take many years to reach this level of connection), and the resulting sewage flows are correspondingly high. Blockages are very rare, even in the upper reaches of the network where the flow is intermittent: here solids progress in a sequence of deposition, transport, deposition, transport; and this continues until the sewer has drained a sufficiently large area for the flow to cease being intermittent. This sequence of deposition–transport–deposition–transport is more efficient in small diameter sewers than in sewers of large diameter.

Simplified sewerage is a sanitation system now widely used in Brazil and elsewhere in Latin America, and it is becoming increasingly used in Africa and Asia (it was introduced to Pakistan in 1985 in Christy Nagar, a very low-income slum area of Karachi). Simplified sewerage, especially its condominial version, is without doubt one of the most important advances ever made in sanitation and, given the extremely high rate of urbanization in many developing countries, in most cases it will be the only technically feasible, economically appropriate and financially affordable sanitation option available for high-density, low-income areas (see Chapter 13).

9.3 HYDRAULIC DESIGN

Simplified sewer design is based either on the flow of sewage achieving a *minimum self-cleansing velocity* at peak flow, or on a *minimum tractive tension* being achieved at peak flow, which ensures solids transport along the sewer. The latter is the more modern approach, and it is incorporated in the current Brazilian sewerage design code.

Prior to presenting these two design approaches, it is necessary to be familiar with the geometric properties of a circular section, since almost all sewers used in simplified sewerage are circular in section.

9.3.1 Properties of a circular section

Unlike settled sewerage, in which the flow of sewage can vary from open channel flow to pressure flow and back again to open channel flow, the flow in simplified sewers is always open channel flow. Manning's equation, or Macedo's modification of it (subsection 9.3.2), is normally used, and this requires knowledge of the area of flow and the hydraulic radius (= the area of flow divided by the wetted perimeter). Both these parameters vary with the depth of flow, as shown in Figure 9.3. From this figure, the following trigonometric relationships can be derived for the area of flow (a, m^2), the wetted perimeter (p, m), the hydraulic radius (r, m) and the

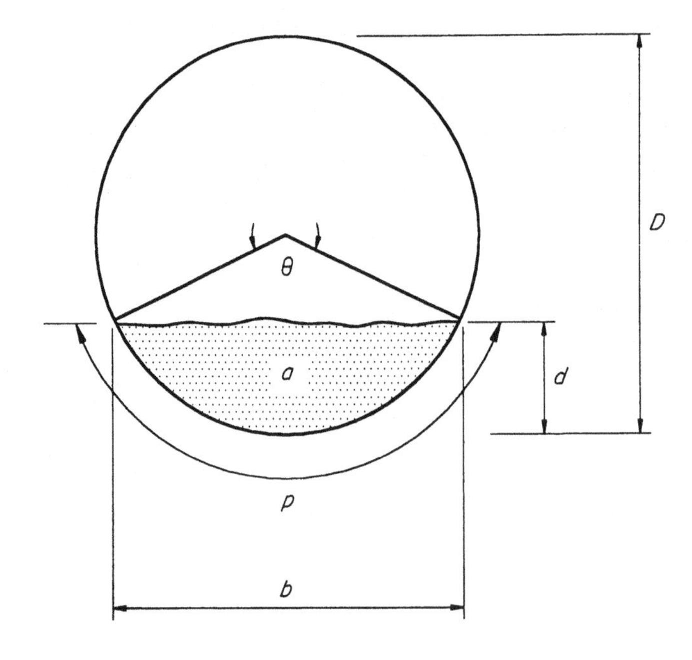

Figure 9.3 Definition of parameters for open channel flow in a circular sewer

breadth of flow (b, m) (b is used in consideration of the risk of hydrogen sulphide generation—subsection 9.3.5):

$$\theta = 2\cos^{-1}[1 - 2(d/D)] \quad (9.1)$$

where θ = angle of flow (as shown in Figure 9.1), radians
d = depth of flow, m
D = pipe diameter, m

The ratio d/D is termed the proportional depth of flow.

In equations (9.1)–(9.12) θ is measured in *radians*. If θ is expressed in degrees, then θ in these equations should be replaced by $(2\pi\theta/360)$, since 2π radians equals 360°.

$$a = D^2[(\theta - \sin\theta)/8] \quad (9.2)$$

$$p = \theta D/2 \quad (9.3)$$

The hydraulic radius (r) is (a/p), and is therefore given by equation (9.2) divided by equation (9.3):

$$r = (D/4)[1 - ((\sin\theta)/\theta)] \quad (9.4)$$

$$b = D\sin(\theta/2) \quad (9.5)$$

Of course, when $d = D$ (i.e. when the pipe is flowing just full), then $a = A = \pi D^2/4$; $p = P = \pi D$; and $r = R = D/4$.

Sometimes reference is made to the proportional area (a/A) and the proportional hydraulic radius (r/R):

$$a/A = (\theta - \sin\theta)/2\pi \quad (9.6)$$

$$p/P = \theta/2\pi \quad (9.7)$$

$$r/R = 1 - [(\sin\theta)/\theta] \quad (9.8)$$

Alternatively, the following equations may be used:

$$a = k_a D^2 \quad (9.9)$$

$$r = k_r D \quad (9.10)$$

where, from equations (9.2) and (9.3):

$$k_a = \tfrac{1}{8}(\theta - \sin\theta) \quad (9.11)$$

$$k_r = \tfrac{1}{4}[1 - ((\sin\theta)/\theta)] \quad (9.12)$$

Manning's equation

As noted in subsection 8.3.3, Manning's equation (equation (8.3)) relates the velocity of flow to the slope and the hydraulic radius:

$$v = (1/n)r^{2/3}i^{1/2} \tag{9.13}$$

where v = velocity of flow at d/D, m/s
n = Manning's roughness coefficient.

Since $q = av$:

$$q = (1/n)ar^{2/3}i^{1/2} \tag{9.14}$$

where q = flow at d/D, m^3/s.
When $d = D$, $v = V$, $q = Q$, $a = A$ and $r = R$, so that:

$$V = (1/n)R^{2/3}i^{1/2} \tag{9.15}$$

$$Q = (1/n)AR^{2/3}i^{1/2} \tag{9.16}$$

Dividing equations (9.13) and (9.14) by equations (9.15) and (9.16), respectively, gives the following expressions for the proportional velocity (v/V) and the proportional flow (q/Q):

$$v/V = (r/R)^{2/3} \tag{9.17}$$

$$q/Q = (a/A)(r/R)^{2/3} \tag{9.18}$$

Substituting equations (9.6) and (9.8) for a/A and r/R gives:

$$v/V = [1 - ((\sin\theta)/\theta)]^{2/3} \tag{9.19}$$

$$q/V = [(\theta - \sin\theta)/2\pi]\,[1 - ((\sin\theta)/\theta)]^{2/3} \tag{9.20}$$

Figure 9.4 shows the variations of v/V and q/Q with d/D, from which it can be noted:

- $v = V$ at both $d/D = 0.5$ and $d/D = 1$;
- the maximum velocity occurs at $d/D = 0.81$ and is given by $v/V = 1.14$; and
- the maximum flow occurs at $d/D = 0.94$ and is given by $q/Q = 1.07$.

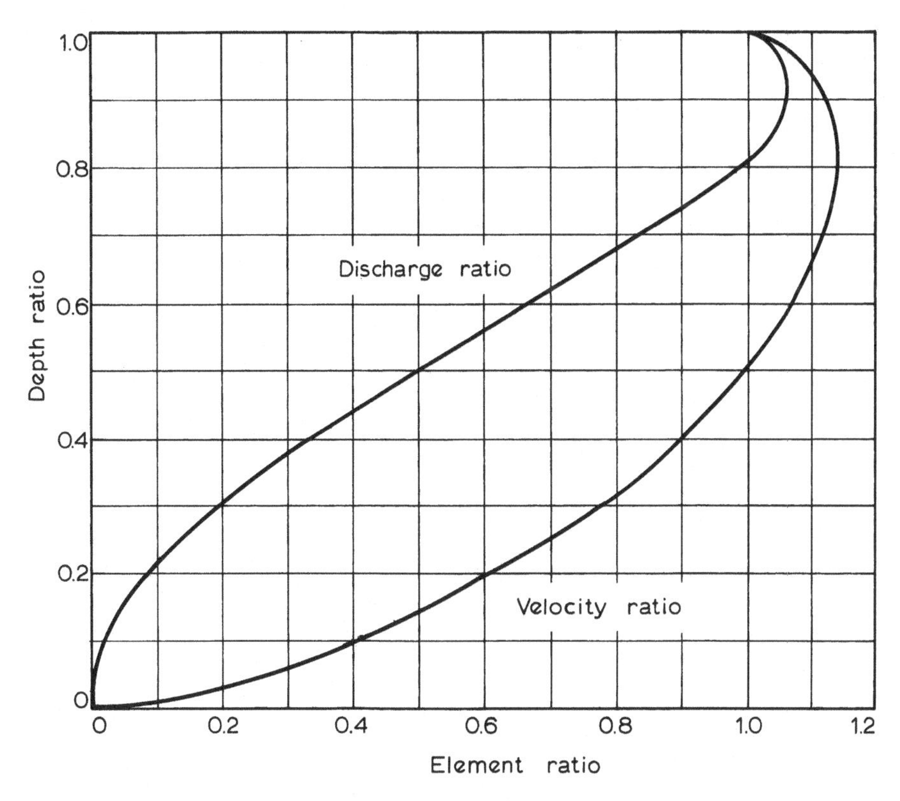

Figure 9.4 Variation of proportional velocity and proportional flow with proportional depth in a circular sewer

Table 9.1 lists, for $0.05 < d/D < 1$, the values of k_a, k_r, a/A, r/R, v/V and q/Q.

9.3.2 Design based on minimum self-cleansing velocity

If sewage flows in a sewer at too low a velocity, there will be a risk of solids deposition and sewer blockage. To prevent this happening, sewers are designed such that the sewage flow achieves a specified minimum self-cleansing velocity at least once a day, at peak flow. In conventional sewerage the required self-cleansing velocity is quite high, > 0.6 m/s and often closer to 1 m/s. In simplified sewerage schemes in peri-urban areas in Brazil, satisfactory sewer performance has been obtained with a self-cleansing velocity of 0.5 m/s being achieved at a daily peak flow equal to 1.8 times the average daily flow.

Table 9.1 Hydraulic elements of a circular section

d/D	k_a	k_r	a/A	r/R	v/V	q/Q
0.02	0.0037	0.0132	0.0048	0.0528	0.1408	0.0007
0.04	0.0105	0.0262	0.0134	0.1047	0.2221	0.0030
0.06	0.0192	0.0389	0.0245	0.1555	0.2892	0.0071
0.08	0.0294	0.0513	0.0375	0.2053	0.3480	0.0130
0.10	0.0409	0.0635	0.0520	0.2541	0.4012	0.0209
0.12	0.0534	0.0755	0.0680	0.3018	0.4500	0.0306
0.14	0.0668	0.0871	0.0851	0.3485	0.4953	0.0421
0.16	0.0811	0.0986	0.1033	0.3942	0.5376	0.0555
0.18	0.0961	0.1097	0.1224	0.4388	0.5775	0.0707
0.20	0.1118	0.1206	0.1424	0.4824	0.6151	0.0876
0.22	0.1281	0.1312	0.1631	0.5248	0.6507	0.1061
0.24	0.1449	0.1416	0.1845	0.5662	0.6844	0.1263
0.26	0.1623	0.1516	0.2066	0.6065	0.7165	0.1480
0.28	0.1800	0.1614	0.2292	0.6457	0.7471	0.1712
0.30	0.1982	0.1709	0.2523	0.6838	0.7761	0.1958
0.32	0.2167	0.1802	0.2758	0.7207	0.8038	0.2217
0.34	0.2355	0.1891	0.2998	0.7565	0.8302	0.2489
0.36	0.2546	0.1978	0.3241	0.7911	0.8554	0.2772
0.38	0.2739	0.2062	0.3486	0.8246	0.8794	0.3066
0.40	0.2934	0.2142	0.3735	0.8569	0.9022	0.3369
0.42	0.3130	0.2220	0.3985	0.8880	0.9239	0.3682
0.44	0.3328	0.2295	0.4237	0.9179	0.9445	0.4002
0.46	0.3527	0.2366	0.4491	0.9465	0.9640	0.4329
0.48	0.3727	0.2435	0.4745	0.9739	0.9825	0.4662
0.50	0.3927	0.2500	0.4999	1.0000	1.0000	0.5000
0.52	0.4127	0.2562	0.5254	1.0248	1.0165	0.5340
0.54	0.4327	0.2621	0.5508	1.0483	1.0319	0.5684
0.56	0.4526	0.2676	0.5761	1.0704	1.0464	0.6029
0.58	0.4724	0.2728	0.6013	1.0912	1.0599	0.6374
0.60	0.4920	0.2776	0.6264	1.1106	1.0724	0.6718
0.62	0.5115	0.2821	0.6512	1.1285	1.0839	0.7059
0.64	0.5308	0.2862	0.6758	1.1449	1.0944	0.7396
0.66	0.5499	0.2900	0.7001	1.1599	1.1039	0.7728
0.68	0.5687	0.2933	0.7240	1.1732	1.1124	0.8054
0.70	0.5872	0.2962	0.7476	1.1849	1.1198	0.8371
0.72	0.6054	0.2987	0.7707	1.1950	1.1261	0.8679
0.74	0.6231	0.3008	0.7933	1.2033	1.1313	0.8975
0.76	0.6405	0.3024	0.8153	1.2097	1.1353	0.9257
0.78	0.6573	0.3036	0.8368	1.2143	1.1382	0.9524
0.80	0.6736	0.3042	0.8575	1.2168	1.1397	0.9773

Table 9.1 *(continued)*

d/D	k_a	k_r	a/A	r/R	v/V	q/Q
0.82	0.6893	0.3043	0.8775	1.2171	1.1399	1.0003
0.84	0.7043	0.3038	0.8966	1.2150	1.1387	1.0209
0.86	0.7186	0.3026	0.9148	1.2104	1.1358	1.0390
0.88	0.7320	0.3007	0.9319	1.2029	1.1311	1.0541
0.90	0.7445	0.2980	0.9478	1.1921	1.1243	1.0657
0.92	0.7560	0.2944	0.9624	1.1775	1.1151	1.0731
0.94	0.7662	0.2895	0.9754	1.1579	1.1027	1.0755
0.96	0.7749	0.2829	0.9865	1.1316	1.0859	1.0712
0.98	0.7816	0.2735	0.9951	1.0941	1.0618	1.0566
1.00	0.7854	0.2500	1.0000	1.0000	1.0000	1.0000

Equation (8.2) is thus rewritten as:

$$q_{\mathrm{h}} = 1.8 \times 10^{-5} pw \tag{9.21}$$

where q_{h} = peak flow from each household, l/s
p = household size
w = water consumption, litres per caput per day, lcd (the equation incorporates a return factor of 0.85—see subsection 8.3.3)

In practice, when $q_{\mathrm{h}} < 2.2$ l/s, it is taken as this value, which is the peak discharge from a flush toilet. This is not, of course, likely as the flow is attenuated in the house connection, but nonetheless it has been found to be quite satisfactory. Current Brazilian practice is to assume a minimum peak flow of 1.5 (rather than 2.2) l/s.

Equation (9.21) can be restated as:

$$q = 1.8 \times 10^{-5} pw \tag{9.22}$$

where q = peak flow in the sewer, l/s
p = contributing population (= number of households served × average household size)

The assumed peak flow of 1.5 l/s would only be achieved by as many as 140 households of six persons with a water consumption of 100 lcd (200 households for 2.2 l/s).

Manning's equation (equation (9.13)) was simplified by

Macedo (a Brazilian sanitary engineer, 1916–1984), as follows: he cubed each side of the equation, and multiplied the left-hand side by v and the right-hand side by q/a (where q is the flow in the sewer, m^3/s and a the area of flow, m^2):

$$v^4 = (q/a)(n^{-3}r^2 i^{3/2}) \tag{9.23}$$

Putting $M = (r^2/a)^{1/4}$, the Macedo–Manning equation becomes:

$$v = Mn^{-3/4}q^{1/4}i^{3/8} \tag{9.24}$$

Since $r^2/a = a/p^2$, equations (9.2) and (9.2) show that:

$$M = [(\theta - \sin\theta)/2\theta^2]^{1/4} \tag{9.25}$$

In fact, for $0.14 < d/D < 0.92$, M is essentially constant and equals 0.61. Taking n as 0.013, the Macedo–Manning equation can be written as:

$$v = 15.8\, q^{1/4}i^{3/8} \tag{9.26}$$

The minimum gradient (I_{min}, m/m) required to achieve a minimum self-cleansing velocity of v_{sc} (m/s) is thus given by:

$$I_{min} = (v_{sc}/15.8)^{8/3}q^{-2/3} \tag{9.27}$$

For $v_{sc} = 0.5$ m/s:

$$I_{min} = 1 \times 10^{-4}q^{-2/3} \tag{9.28}$$

Changing the units of q from m^3/s to l/s gives:

$$I_{min} = 0.01q^{-2/3} \tag{9.29}$$

For the minimum peak flow of 2.2 l/s, the value of I_{min} is 0.006 or 1 in 167, and this gradient has been successfully used in northeast Brazil for the flow in a 100 mm sewer serving 60 households. For a minimum peak flow of 1.5 l/s, I_{min} is 0.0076 or 1 in 131.

Selection of sewer diameter

Using equations (9.9) and (9.10), equation (9.14) can be rewritten as:

$$q = (1/n)k_a D^2 (k_r D)^{2/3} i^{1/2} \qquad (9.30)$$

Rearranging and writing $i = I_{\min}$:

$$D = n^{3/8} k_a{}^{-3/8} k_r{}^{-1/4} (q/I_{\min}^{1/2})^{3/8} \qquad (9.31)$$

Since k_a and k_r are functions of d/D, the sewer diameter depends not only on the flow and the corresponding minimum gradient, but also on d/D. In simplified sewerage the usual limits for d/D are as follows:

$$0.2 < d/D < 0.8$$

The lower limit ensures that there is sufficient velocity to flow to prevent solids depostion in the initial part of the design period, and the upper limit provides for sufficient ventilation at the end of the design period.

The sewer diameter may be determined by the following sequence of calculations:

1. Calculate, using equation (9.22), the initial and final sewage flows (q_i and q_f respectively, l/s), which are the flows occurring at the start and end of the design period. The increase in flow is due either to an increase in population or an increase in water consumption, or both.
2. Calculate $I_{\min}$ from equation (9.29) with $q = q_i$.
3. Calculate D from equation (9.31) using $q = q_f$ (in m^3/s) for $d/D = 0.8$ (i.e. using equations (9.1), (9.11) and (9.12), or Table 9.1). The diameter so calculated is unlikely to be a commercially available size; choose the next larger diameter that is available (i.e. if $D = 76$ mm, say, choose 100 mm).
4. In practice, instead of calculating D in this way, Table 9.2 (in which $n = 0.013$) can be used as follows:
   (a) Calculate $q_f/I_{\min}^{1/2}$ with q_f in m^3/s, and find this value in Table 9.2 where d/D is close to, but less than, 0.8.

Table 9.2 Design chart for simplified sewers based on Manning's equation with $n = 0.013$, and v in m/s, I in m/m, q in m^3/s and the sewer diameter D in mm

d/D	$D = 100$		$D = 150$		$D = 225$		$D = 300$	
	$v/I^{1/2}$	$q/I^{1/2}$	$v/I^{1/2}$	$q/I^{1/2}$	$v/I^{1/2}$	$q/I^{1/2}$	$v/I^{1/2}$	$q/I^{1/2}$
0.02	0.9260	0.0000	1.2135	0.0001	1.5901	0.0003	1.9262	0.0006
0.04	1.4607	0.0002	1.9140	0.0005	2.5081	0.0013	3.0383	0.0029
0.06	1.9017	0.0004	2.4920	0.0011	3.2654	0.0032	3.9558	0.0068
0.08	2.2888	0.0007	2.9992	0.0020	3.9300	0.0059	4.7609	0.0126
0.10	2.6383	0.0011	3.4572	0.0032	4.5302	0.0094	5.4880	0.0202
0.12	2.9593	0.0016	3.8778	0.0047	5.0814	0.0137	6.1557	0.0296
0.14	3.2573	0.0022	4.2683	0.0064	5.5930	0.0189	6.7754	0.0408
0.16	3.5359	0.0029	4.6334	0.0085	6.0714	0.0249	7.3550	0.0537
0.18	3.7979	0.0037	4.9766	0.0108	6.5212	0.0317	7.8999	0.0684
0.20	4.0451	0.0045	5.3006	0.0133	6.9458	0.0393	8.4142	0.0847
0.22	4.2792	0.0055	5.6074	0.0162	7.3477	0.0477	8.9012	0.1026
0.24	4.5013	0.0065	5.8984	0.0192	7.7291	0.0567	9.3631	0.1221
0.26	4.7124	0.0076	6.1750	0.0225	8.0915	0.0665	9.8022	0.1431
0.28	4.9132	0.0088	6.4382	0.0261	8.4364	0.0769	10.2200	0.1656
0.30	5.1045	0.0101	6.6888	0.0298	8.7648	0.0879	10.6178	0.1894
0.32	5.2867	0.0115	6.9276	0.0338	9.0777	0.0996	10.9968	0.2144
0.34	5.4604	0.0129	7.1551	0.0379	9.3759	0.1118	11.3580	0.2407
0.36	5.6258	0.0143	7.3719	0.0422	9.6599	0.1245	11.7022	0.2681
0.38	5.7834	0.0158	7.5784	0.0467	9.9305	0.1377	12.0300	0.2965
0.40	5.9334	0.0174	7.7750	0.0513	10.1881	0.1513	12.3420	0.3259
0.42	6.0761	0.0190	7.9619	0.0561	10.4331	0.1653	12.6388	0.3561
0.44	6.2116	0.0207	8.1395	0.0610	10.6658	0.1797	12.9206	0.3870
0.46	6.3401	0.0224	8.3079	0.0659	10.8865	0.1944	13.1880	0.4187
0.48	6.4618	0.0241	8.4674	0.0717	11.0955	0.2094	13.4412	0.4509
0.50	6.5768	0.0258	8.6181	0.0761	11.2929	0.2245	13.6804	0.4835
0.52	6.6852	0.0276	8.7601	0.0813	11.4789	0.2398	13.9057	0.5165
0.54	6.7870	0.0294	8.8934	0.0866	11.6537	0.2553	14.1174	0.5497
0.56	6.8822	0.0311	9.0182	0.0918	11.8172	0.2707	14.3155	0.5831
0.58	6.9709	0.0329	9.1345	0.0971	11.9696	0.2862	14.5001	0.6164
0.60	7.0531	0.0347	9.2422	0.1023	12.1107	0.3017	14.6711	0.6497
0.62	7.1288	0.0365	9.3414	0.1075	12.2407	0.3170	14.8285	0.6827
0.64	7.1979	0.0382	9.4319	0.1127	12.3593	0.3321	14.9722	0.7153
0.66	7.2603	0.0399	9.5137	0.1177	12.4664	0.3471	15.1020	0.7474
0.68	7.3159	0.0416	9.5865	0.1227	12.5619	0.3617	15.2177	0.7789
0.70	7.3646	0.0432	9.6503	0.1275	12.6455	0.3759	15.3189	0.8096
0.72	7.4061	0.0448	9.7048	0.1322	12.7169	0.3897	15.4054	0.8393
0.74	7.4404	0.0464	9.7497	0.1367	12.7757	0.4030	15.4766	0.8680
0.76	7.4670	0.0478	9.7845	0.1410	12.8214	0.4157	15.5320	0.8953
0.78	7.4856	0.0492	9.8090	0.1451	12.8534	0.4277	15.5708	0.9211
0.80	7.4959	0.0505	9.8224	0.1489	12.8710	0.4389	15.5921	0.9452

Table 9.2 (*continued*)

d/D	$D = 100$		$D = 150$		$D = 225$		$D = 300$	
	$v/I^{1/2}$	$q/I^{1/2}$	$v/I^{1/2}$	$q/I^{1/2}$	$v/I^{1/2}$	$q/I^{1/2}$	$v/I^{1/2}$	$q/I^{1/2}$
0.82	7.4972	0.0517	9.8241	0.1524	12.8732	0.4492	15.5947	0.9674
0.84	7.4888	0.0527	9.8131	0.1555	12.8588	0.4585	15.5773	0.9874
0.86	7.4698	0.0537	9.7882	0.1583	12.8261	0.4666	15.5377	1.0048
0.88	7.4389	0.0545	9.7477	0.1605	12.7731	0.4733	15.4735	1.0194
0.90	7.3944	0.0551	9.6894	0.1623	12.6967	0.4786	15.3810	1.0306
0.92	7.3336	0.0554	9.6098	0.1635	12.5923	0.4819	15.2545	1.0379
0.94	7.2522	0.0556	9.5031	0.1638	12.4526	0.4830	15.0852	1.0402
0.96	7.1421	0.0553	9.3588	0.1632	12.2634	0.4811	14.8561	1.0360
0.98	6.9830	0.0546	9.1504	0.1609	11.9904	0.4745	14.5253	1.0218
1.00	6.5768	0.0517	8.6181	0.1523	11.2929	0.4490	13.6804	0.9670

The sewer diameter D is given at the top of the column in which $q_f/I_{min}^{1/2}$ is found.

(b) Read the corresponding values of $v/I^{1/2}$ from Table 9.2 and so calculate v_f. This step is based on a rearrangement of equation (9.13), which incorporates equation (9.10):

$$v/I^{1/2} = (1/n)\,(k_r D)^{2/3} \tag{9.32}$$

(c) Calculate $q_i/I_{min}^{1/2}$ with q_i in m^3/s; find this value in the same column in Table 9.2 as used in step (a), and read the corresponding value of d/D and calculate v_i from the corresponding value of $v/I^{1/2}$.

Use of Table 9.2 thus permits easy determination not only of the sewer diameter, but also of the velocities and proportional depths of flow at the beginning and end of the design period. A worked example is given in section 9.7.

9.3.3 Design based on minimum tractive tension

Tractive tension (or boundary shear stress) is the tangential force exerted by the flow of sewage per unit of wetted boundary area. It is denoted by the symbol τ and has units of N/m^2 (i.e. Pascals, Pa). As shown in Figure 9.5, and considering a

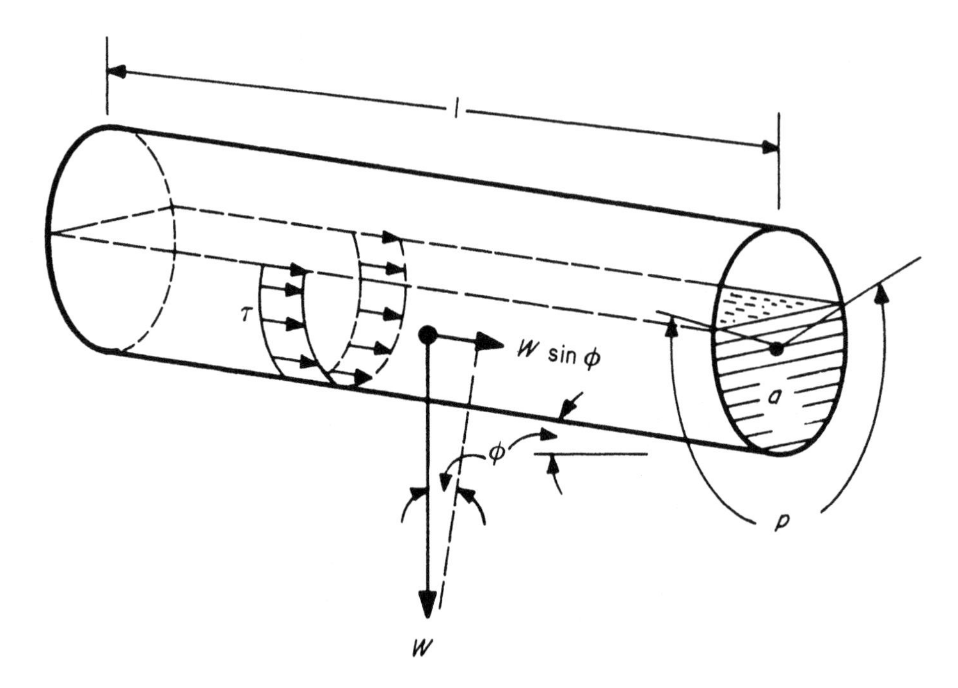

Figure 9.5 Definition of parameters for tractive tension in a circular sewer

mass of sewage of length l m and cross-sectional area a m^2, which has a wetted perimeter of p m, the tractive tension is given by the component of the weight (W, Newtons) of this mass of sewage in the direction of flow divided by its corresponding wetted boundary area (i.e. the area in which it is in contact with the sewer = pl):

$$\tau = W \sin \phi / pl \tag{9.33}$$

The weight W is given by:

$$W = \rho g a l \tag{9.34}$$

so that, since a/p is the hydraulic radius r:

$$\tau = \rho g r \sin \phi \tag{9.35}$$

where ρ = density of sewage, kg/m^3
g = acceleration due to gravity, m/s^2

When ϕ is small, $\sin \phi = \tan \phi$, and $\tan \phi$ is the sewer gradient, i (m/m). Thus, equation (9.35) can be rewritten as:

$$\tau = \rho gri \tag{9.36}$$

Using equation (9.10) and rearranging:

$$D = (\tau/\rho g)/k_r i \tag{9.37}$$

Substituting this expression for D in equation (9.30):

$$q = (1/n)k_a D^2(k_r D)^{2/3} i^{1/2} \tag{9.30}$$

i.e.

$$= (1/n)k_a k_r^{2/3} D^{8/3} i^{1/2}$$

$$= (1/n)k_a k_r^{2/3}[(\tau/\rho g)/k_r i]^{8/3} i^{1/2}$$

$$q = (1/n)k_a k_r^{-2}(\tau/\rho g)^{8/3} i^{-13/6} \tag{9.38}$$

The minimum gradient, I_{min} is given by rearranging this equation with $\tau = \tau_{min}$ at $d/D = 0.2$ (i.e. $k_a = 0.1118$ and $k_r = 0.1206$ from Table 9.1), and with $\rho = 1000$ kg/m^3, $g = 9.81$ m/s^2 and $n = 0.013$, as follows:

$$I_{min} = [(1/n)k_a k_r^{-2}]^{6/13}[\tau_{min}/\rho g]^{16/13} q^{-6/13} \tag{9.39}$$

$$I_{min} = 2.33 \times 10^{-4}(\tau_{min})^{16/13} q^{-6/13} \tag{9.40}$$

Equation (9.39) is the minimum tractive tension design equation, equivalent to equation (9.27) in the minimum self-cleansing velocity design approach. For simplified sewerage, a satisfactory value of τ_{min} is 1 Pa; thus:

$$I_{min} = 2.33 \times 10^{-4} q^{-6/13} \tag{9.41}$$

Changing the units of q from m^3/s to l/s gives:

$$I_{min} = 5.64 \times 10^{-3} q^{-6/13} \tag{9.42}$$

Selection of sewer diameter

The procedure for the selection of sewer diameter is exactly the same as that given in subsection 9.3.2, except that equation (9.42) is used instead of equation (9.29). A worked design example is given in section 9.7.

Comparison of design approaches

Equations (9.29) and (9.42) can be used to obtain the minimum gradients required for the minimum flows of 2.2 and 1.5 l/s, as follows:

Minimum flow:	2.2 l/s	1.5 l/s
Minimum velocity = 0.5 m/s:	1 in 167	1 in 131
Minimum tractive tension = 1 Pa:	1 in 255	1 in 214

Thus, the minimum tractive tension design approach is more economical than the minimum self-cleansing velocity design approach: it results in significantly reduced gradients, hence reduced excavation and thus reduced costs. There is considerable experience from southern Brazil of the long-term satisfactory operation of simplified sewers designed with a minimum tractive tension of 1 Pa, and thus the use of this design method is to be preferred to that based on minimum self-cleansing velocity.

Indeed, so successful has this experience been that SANEPAR, the water and sewerage company of the southern Brazilian state of Paraná, is now beginning to design simplified sewerage schemes with reduced values of τ_{min} and Manning's n of 0.59 Pa and 0.010, respectively. (This value of 0.59 Pa can be derived from equation (9.36) with $r = 0.1206D$ for $d/D = 0.2$ and with $i = 1$ in $2000D$.) This results in even shallower gradients, as calculated from equation (9.39) (with q in m^3/s) for $d/D = 0.2$:

	2.2 l/s	1.5 l/s
$\tau_{min} = 1$ Pa, $n = 0.013$:	1 in 255	1 in 214
$\tau_{min} = 0.59$ Pa, $n = 0.010$:	1 in 433	1 in 363

Operation of simplified sewers designed with these reduced values of τ_{min} and n appears to be satisfactory, but general adoption of these values must await long-term operational experience.

9.3.4 Hydrogen sulphide control

Hydrogen sulphide control is important in concrete and asbestos–cement sewers, otherwise corrosion of the sewer crown by microbially produced sulphuric acid can occur (Figure 9.6). Although vitrified clay and PVC pipes, which are normally used for the small diameter sewers in simplified sewerage, are immune to such corrosion, the following method due to Pomeroy for predicting sulphide attack is given both for completeness and for when the use of concrete or asbestos–cement sewers is unavoidable. The Pomeroy 'Z factor' is defined as:

$$Z = 3\,(BOD_5)(1.07)^{T-20} i^{-1/2} q^{-1/3} (p/b) \qquad (9.43)$$

where BOD_5 = 5-day, 20°C biochemical oxygen demand of the sewage, mg/l
T = temperature, °C
i = sewer gradient, m/m
q = sewage flow, l/s
p = wetted perimeter, m
b = breadth of flow (see Figure 9.3), m

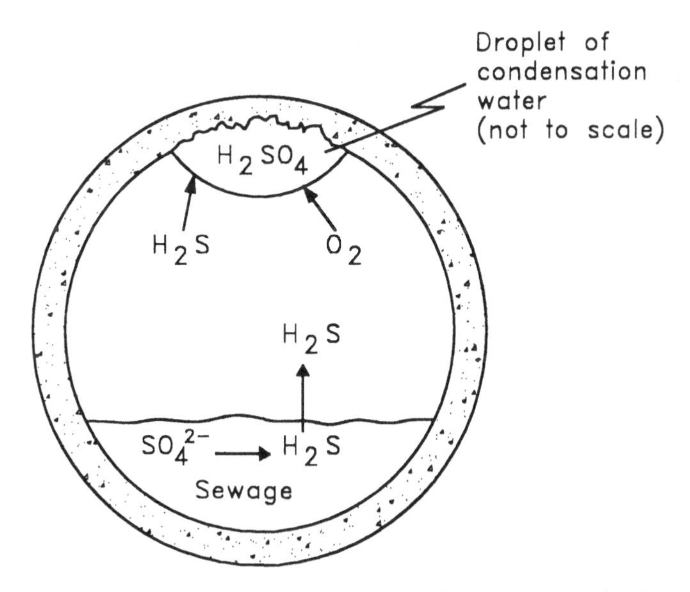

Figure 9.6 Microbially induced corrosion of the crown of concrete or asbestos cement sewers: sulphates in the sewage are reduced anaerobically by sulphate-reducing bacteria to hydrogen sulphide, some of which leaves the sewage to raise its partial pressure in the atmosphere above the flow (Henry's law), and then some of this H_2S goes into solution (Henry's law again) in droplets of condensation water clinging to the sewer crown—this H_2S is oxidized by the aerobic bacterium *Thiobacillus thioparus* to sulphuric acid (H_2SO_4), which corrodes the concrete. Sewer crown collapse within 10–20 years is not uncommon

and 3 is the conversion factor resulting from changing the units of q from ft^3/s in Pomeroy's original equation to l/s.

The value of Z calculated from equation (9.43) can be used diagnostically as follows:

$$Z < 5000\text{: } H_2S \text{ generation unlikely}$$

$$5000 < Z < 10\,000\text{: } H_2S \text{ generation possible}$$

$$Z > 10\,000\text{: } H_2S \text{ generation very likely}$$

With simplified sewerage, hydrogen sulphide generation can be expected to be a common problem. For example, for a flow of 1.5 l/s of sewage with a BOD_5 of 250 mg/l at 25°C in a sewer laid at 1 in 214 and flowing at a proportional depth of flow of 0.2, Z can be calculated as follows, using equations (9.1), (9.3) and (9.5) to calculate p/b for $d/D = 0.2$:

$$\theta/2 = \cos^{-1}[1 - 2\,(d/D)]$$

$$= 0.927 \text{ radian}$$

$$p/b = (\theta/2)/\sin(\theta/2)$$

$$= 1.159$$

$$Z = 3 \times 250(1.07)^5(1/214)^{-1/2}(1.5)^{-1/3}(1.159)$$

$$= 16\,000$$

This is why the small diameter pipes used in simplified sewerage schemes are normally of either vitrified clay or PVC.

9.3.5 Sewer gradient and ground slope

The slope of the ground surface (S, m/m) may be (a) less than, (b) equal to, (c) greater than, or (d) much greater than, the minimum sewer gradient (I_{min}, m/m) calculated from either equation (9.29) or equation (9.42). Furthermore, the depth to the invert of the upstream end of the length of sewer under consideration may be (a) equal to, or (b) greater than, the minimum depth permitted (h_{min}, m), which is given by:

$$h_{min} = C + D \tag{9.44}$$

where C = minimum required cover, m (see Figure 9.7)
D = sewer diameter, m

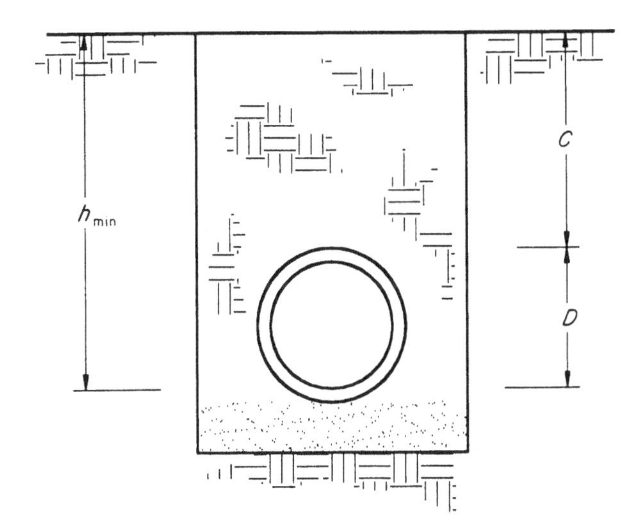

Figure 9.7 The minimum depth (h_{min}) to which a sewer is laid is the sum of the minimum depth of cover (C) and the sewer diameter (D)

There are six cases likely to be encountered in practice. These are (see also Figure 9.8):

Case 1. $S < I_{min}$ and the invert depth of the upstream end of the sewer (h_1, m) $\geqslant h_{min}$: choose $i = I_{min}$ and calculate the invert depth of the downstream end of the sewer (h_2, m) as:

$$h_2 = h_1 + (I_{min} - S)L \qquad (9.45)$$

where L = length of sewer under consideration, m.

Case 2. $S = I_{min}$ and $h_1 \geqslant h_{min}$: choose $i = I_{min}$ and $h_2 = h_1$.

Case 3. $S > I_{min}$ and $h_1 = h_{min}$: choose $i = S$ and $h_2 = h_1$.

Case 4. $S > I_{min}$ and $h_1 > h_{min}$: choose $h_2 = h_{min}$ and calculate the sewer gradient from:

$$i = S + (h_{min} - h_1)/L \qquad (9.46)$$

subject to $i \nless I_{min}$.

Case 5. $S > I_{min}$ and $h_1 > h_{min}$, as in Case 4, but an alternative solution is to choose $i = I_{min}$ and calculate h_2 from equation (9.45). The choice between these alternative solutions is made on the basis of minimum excavation.

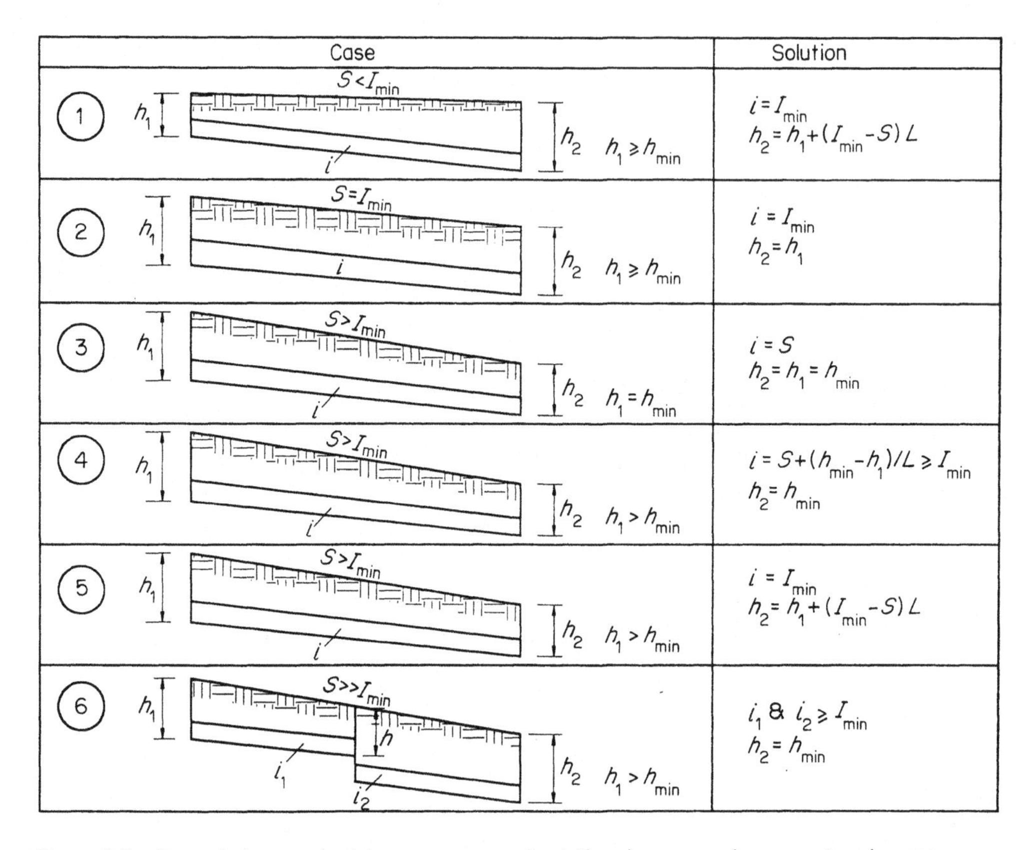

Figure 9.8 Ground slope and minimum sewer gradient: the six commonly encountered cases

Case 6. $S \gg I_{\min}$ and $h_1 \geqslant h_{\min}$: here, it is usually sensible to divide L into two or more substretches with $h_2 = h_{\min}$ and $i \ll S$ (but obviously $\geqslant I_{\min}$) in order to minimize excavation. A drop manhole is placed at the substretch junction.

9.3.6 Number of households served

Suppose sewer pipes are available in 100, 150, 225 and 300 mm diameter. What is the maximum number of households that can be served by each size sewer?

Take, for example, the household size to be five persons, each with a water consumption of 100 l/day. The peak sewage flow, assuming a return factor of 0.85, is given by equation (9.21) as:

$$q_h = 1.8 \times 10^{-5} \times 5 \times 100$$
$$= 0.009 \text{ l/s per household}$$

Suppose that the low-income housing area is fully developed—i.e. there is no room for any further houses. Thus, the only future increase in sewage flow will be due to an increase in water consumption. Examination of Table 9.1 shows that for:

$$d/D = 0.6\text{: } q/Q = 0.6718$$

and

$$\text{for } d/D = 0.8\text{: } q/Q = 0.9773$$

So designing the sewers to flow now with a d/D of 0.6 allows for an increase in water consumption to:

$$100\ (0.9773/0.6718) = 146 \text{ lcd}$$

which is more than adequate.

Equations (9.39) and (9.31) are now solved for $d/D = 0.6$ (i.e. using Table 9.1, $k_a = 0.4920$ and $k_r = 0.2776$), with $\tau = 1$ Pa and expressing q in l/s, as follows:

$$I_{min} = 0.00518 q^{-6/13} \qquad (9.47)$$
$$D = 0.0264 (q/I_{min}^{1/2})^{3/8} \qquad (9.48)$$

Substituting equation (9.47) into equation (9.48) and rearranging:

$$q = 310 D^{13/6} \qquad (9.49)$$

and changing the units of D from metres to millimetres gives:

$$q = 9.8 \times 10^{-5} D^{13/6} \qquad (9.50)$$

Now the peak flow per household is 0.009 l/s, so that q given by:

$$q = 0.009 N \qquad (9.51)$$

where N = number of households served. Thus:

$$N = 10.89 \times 10^{-3} D^{13/6} \qquad (9.52)$$

Thus, a 100 mm diameter sewer can serve up to 234 households; values for the other diameters are given in Table 9.3.

Equation (9.52) is, of course, only valid for the values assumed—i.e. a fully developed housing area, a household size of five, a water consumption of 100 lcd, a return factor of 0.85, a peak factor of 1.8, a Manning's n of 0.013, a minimum tractive tension of 1 Pa and a proportional depth of flow of 0.6. For other values, an equation similar to equation (9.52) has to be derived in the manner shown above (see Design Example 3 in section 9.7).

Simplified design

Table 9.3 and equation (9.47) can be used directly for the design of each stretch of sewer in a fully developed housing area, as follows:

- determine the number of households to be sewered;
- select, from Table 9.3, the sewer diameter (Table 9.3 is based on equation (9.52), which contains the inherent values of the design parameters given above, and so is only valid for these values; for any other value(s) equation (9.52) must be rederived and Table 9.3 recalculated for the desired range of sewer diameter);
- determine the peak flow from equation (9.51) but, if $q < 2.2$ l/s, then take q as 2.2 l/s (subsection 9.3.2); and

Table 9.3 Maximum number of households served by simplified sewers of 100–300 mm diameter[a]

Sewer diameter (mm)	Maximum number of households served
100	234
150	565
225	1360
300	2536

[a]Based on equation (9.52) (see text for inherent values of parameters, such as household size and per caput water consumption, assumed).

- calculate the minimum sewer gradient from equation (9.47), and ensure that the actual sewer gradient is not less than this (subsection 9.3.5).

Sewers so designed have an additional flow capacity of 46 percent, and this should not be exceeded, so that the final d/D is $\not> 0.8$. (If, for any reason, the flow does increase such that d/D will be > 0.8, then consideration should be given to the installation of water-saving plumbing fixtures to reduce water consumption and hence sewage flow.)

9.4 CONSTRUCTION AND MAINTENANCE

9.4.1 Sewer appurtenances

Kitchen sullage should discharge into a standard design grease trap, which should be cleaned out regularly by the householder. Simple junction boxes are used to connect house connections to simplified sewers, or one simplified sewer to another (Figure 9.9). Manholes should be avoided wherever possible: they are expensive, and serve as entry points for grit and infiltration water. Moreover, surveys done in southern Brazil have shown that only very few manholes are ever used—over 95 percent of them are never opened. As with settled sewerage, manholes can be replaced in almost all situations by simple inspection and cleanout units (see subsection 8.4.1 and Figure 8.2).

9.4.2 Sewage treatment

A simplified sewer network can join a conventional sewer if one is conveniently located nearby, or it can discharge into its own sewage treatment works (Chapter 10).

9.4.3 Operation and maintenance

The local sewerage authority has to be responsible for the operation and maintenance of the simplified sewerage network, although in the case of condominial sewerage, maintenance of the in-block sewers can be the responsibility of the

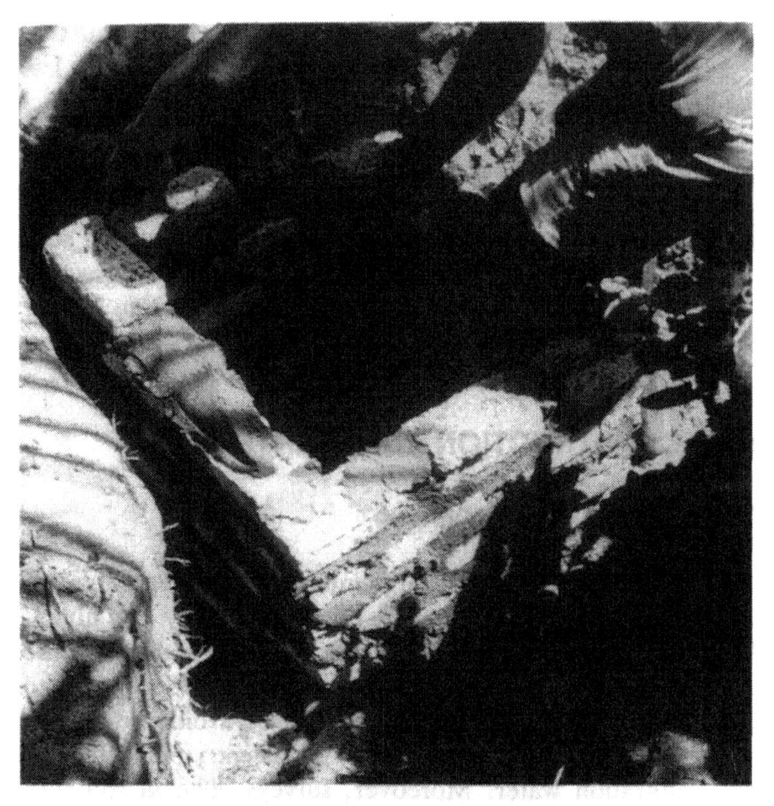

Figure 9.9 Simple junction boxes used to connect house connections to simplified sewers, or one simplified sewer to another —(a) brick juction box in northeast Brazil; (b) concrete pipe junction box in Guatamala

block residents. This reduces costs (see section 9.5), but must be carefully discussed with the residents prior to construction (Figure 9.10). This has worked very well in northeast Brazil, with each household being reponsible for the maintenance of the length of sewer laid in its plot; blockages have been extremely rare, and in-block sewers have been essentially problem-free in many cases for over 15 years. However, if the residents are not fully aware of their maintenance responsibilities, problems can occur, so good liaison has to exist between the sewerage authority and the residents.

9.5 COSTS

Simplified sewerage *is* low-cost. As shown in Figure 9.2, it can even be cheaper than on-site systems such as VIP latrines or

Figure 9.10 An engineer from the Rio Grande do Norte water and sewerage company in Natal, northeast Brazil explaining to local residents how the simplified sewerage system works and what their maintenance responsibilities will be

pour-flush toilets. This depends on the population density; in the low-income areas of Natal in northeast Brazil, where simplified sewerage was developed, it became cheaper than on-site systems at a population density of 160 people per ha. Capital costs were US$325 per household ($65 per caput) (see subsection 12.2 for details of cost recovery). Recent constuction costs reported by the World Bank and the Pan American Health Organization are:

Condominial sewerage in northeast Brazil: US$65–105 per caput

Simplified sewerage in southern Brazil: US$170–240 per caput

Conventional sewerage in southern Brazil: US$240–390 per caput

As would be expected, the condominial sewerage is significantly cheaper than the non-condominial version of simplified sewerage, and of course much cheaper than conventional sewerage.

9.6 FURTHER READING

J. C. Ackers, D. Butler and R. W. P. May, *The Design of Sewers to Control Sediment Problems*. Funders Report No. CP/27. Construction Industry Research and Information Association, London (1994).

Alternative Sewerage Systems. Manual of Practice No. FD-12. Water Pollution Control Federation, Alexandria, VA (1986).

A. Bakalian, A. Wright, R. Otis and J. de Azevedo Netto, *Simplified Sewerage: Design Guidelines*. Water and Sanitation Report No. 7. The World Bank, Washington, DC (1994).

D. Butler and B. R. C. Pinkerton, *Gravity Flow Pipe Design Charts*. Thomas Telford, London (1987).

S. E. Cynamon, *Sîstema Não Convencional de Esgoto Sanitário a Custo Reduzido para Pequenas Coletividades e Áreas Periféricas* (2nd edition). Fundação Oswaldo Cruz, Rio de Janeiro (1986).

C. O. de Andrade Neto, 'Uma solução eficaz e de baixo custo para o esgotamento sanitário urbano'. *Engenharia Sanitária* (Rio de Janeiro) **24** (2), 239–241 (1985).

J. M. de Azevedo Netto, 'A evolução dos sistemas de esgotos'. *Engenharia Sanitária* (Rio de Janeiro) **22** (2), 226–228 (1983).

J. M. de Azevedo Netto, 'Saneamento viável e acessível'. *Engenharia Sanitária* (Rio de Janeiro) **23** (3), 289–300 (1984).

J. M. de Azevedo Netto, *Innovative and Low Cost Technologies Utilized in Sewerage*. Environmental Health Program Technical Series No. 29.

Pan American Health Organization, Washington, DC (1992).
A. S. P. Guimarães, *Redes de Esgotos Simplificadas*, Programa das Nações Unidas para o Desenvolvimento/Ministério do Desenvolvimento Urbano e Meio Ambiente, Brasília (1986).
J. C. O. Machado Neto and M. T. Tsutiya, 'Tensão trativa: um critério econômico para o dimensionamento das tubulações de esgoto'. *Revista DAE* (São Paulo) **45** (140), 73–87 (1985).
D. D. Mara, *The Conservation of Drinking Water Supplies: Techniques for Low-income Settlements*. United Nations Centre for Human Settlements, Nairobi (1989).
D. D. Mara (ed.), *Low-cost Sewerage*. John Wiley & Sons, Chichester (1996, forthcoming).
M. J. Marriott, 'Self-cleansing sewer gradients'. *Journal of the Institution of Water and Environmental Management* **8** (4), 360–361 (1994).
S. R. Mendonça, *Tópicos Advançados em Sistemas de Esgotos Sanitários*. Associaçâo Brasileira de Engenharia Sanitária e Ambiental, Rio de Janeiro (1987).
R. D. Pomeroy, *The Problem of Hydrogen Sulphide in Sewers*, 2nd edition. Clay Pipe Development Association, London (1990).
Projeto de Redes Coletoras de Esgoto Sanitário. Brazilian Standard No. 9649. Associação Brasileira de Normas Técnicas, Rio de Janeiro, RJ (1986).
R. A. Reed, *Sustainable Sewerage: Guidelines for Community Schemes*. IT Publications, London (1995).
E. B. Rondon, *A Critical Evaluation of Shallow Sewerage Systems: A Case Study in Cuiabá, Brazil*. MSc(Eng) Dissertation. University of Leeds (Department of Civil Engineering), Leeds (1990).
J. C. Rodrigues de Melo, 'Sistemas condominiais de esgotos'. *Engenharia Sanitária* (Rio de Janeiro) **24** (2), 237–238 (1985).
G. S. Sinnatamby, *Low-cost Sanitation Systems for Urban Peripheral Areas in Northeast Brazil*. PhD Thesis. University of Leeds (Department of Civil Engineering), Leeds (1983).
G. S. Sinnatamby, *The Design of Shallow Sewer Systems*. United Nations Centre for Human Settlements, Nairobi (1986).
G. Sinnatamby, D. Mara and M. McGarry, 'Sewerage: shallow systems offer hope to slums'. *World Water* **9**, 39–41 (1986).
A. Steer and J. Briscoe, 'New approaches to sanitation: a process of structural learning'. *Ambio* **22** (7), 456–459 (1993).
M. A. Verbanck, R. M. Ashley and A. Bachoc, 'International workshop on origin, occurrence and behaviour of sediments in sewer systems: Summary of conclusions'. *Water Research* **28** (1), 187–194 (1994).
M. Vines and R. Reed, 'Low-cost unconventional sewerage'. *Waterlines* **9** (1), 26–29 (1990).
G. Watson, *Good Sewers Cheap? Agency–Customer Interactions in Low-Cost Urban Sanitation in Brazil*. Water and Sanitation Currents. The World Bank, Washington, DC (1995).

Videotape

The World Bank (address in section 1.5) has produced a 20-minute videotape on the simplified sewerage scheme in the peri-urban area of Orangi in Karachi, Pakistan with the title *Orangi: Streets of Hope*.

9.7. DESIGN EXAMPLES

9.7.1 Example 1

Design a sewer to take the sewage from several housing blocks with a current population of 6000, which is expected to grow to 11 000 in 20 years' time. The water consumption is 100 litres per caput per day (lcd), the return factor 85 percent and the wastewater peak flow factor 1.8. The topography is flat. Assume no infiltration.

Minimum tractive tension design

The initial and future flows are given by:

$$q_i = 100 \times 6000 \times 0.85 \times 1.8$$
$$= 918\,000 \text{ l/day}$$
$$= 10.6 \text{ l/s}$$
$$q_f = 100 \times 11\,000 \times 0.85 \times 1.8$$
$$= 1\,683\,000 \text{ l/d}$$
$$= 19.5 \text{ l/s}$$

Calculate I_{min} from equation (9.42), with $q = q_i$:

$$I_{min} = 5.64 \times 10^{-3} q_i^{-6/13}$$
$$= 5.64 \times 10^{-3}(10.6)^{-6/13}$$
$$= 0.0019 \text{ m/m (1 in 527)}$$

Calculate $q_f/I_{min}^{1/2}$ (with q_f in m^3/s):

$$= 19.5 \times 10^{-3}/0.0019^{1/2}$$
$$= 0.4474$$

From Table 9.2 select a sewer of diameter 225 mm at $d/D = 0.82$. The corresponding value of $v_f/I_{min}^{0.5}$ is 12.8732, from which $v_f = 0.56$ m/s.

Calculate $q_i/I_{min}^{1/2}$ (with q_i in m^3/s):

$$= 10.6 \times 10^{-3}/0.0019^{1/2}$$
$$= 0.2432$$

From Table 9.2 the corresponding value of $v_i/I_{min}^{1/2}$ (in the column under $D = 225$ mm) is 11.4789 for $d/D = 0.52$, from which $v_i = 0.50$ l/s.

Minimum self-cleansing velocity design

Calculate I_{min} from equation (9.29), with $q = q_i$:

$$\begin{aligned} I_{min} &= 0.01 q_i^{-2/3} \\ &= 0.01(10.6)^{-2/3} \\ &= 0.0021 \text{ m/m (1 in 483)} \\ q_f/I_{min}^{1/2} &= 19.5 \times 10^{-3}/0.0021^{1/2} \\ &= 0.4255 \end{aligned}$$

From Table 9.2 choose $D = 225$ mm at $d/D = 0.78$. So $V_f/I_{min}^{1/2} = 12.8534$, from which $v_f = 0.59$ m/s. Now:

$$\begin{aligned} q_i/I_{min}^{1/2} &= 10.6 \times 10^{-3}/0.0021^{1/2} \\ &= 0.2313 \end{aligned}$$

The corresponding value of $v_i/I_{min}^{1/2}$ is 11.4789 at $d/D = 0.52$, from which $v_i = 0.53$ m/s.

Comparison of designs

The two designs, based on the minimum tractive tension and minimum self-cleansing velocity design methods, can be compared as follows:

	Minimum tractive tension (1 Pa)	Minimum self-cleansing velocity (0.5 m/s)
Diameter (mm)	225	225
Gradient	1 in 527	1 in 483
Initial velocity (m/s)	0.50 at $d/D = 0.52$	0.53 at $d/D = 0.52$
Final velocity (m/s)	0.56 at $d/D = 0.82$	0.59 at $d/D = 0.78$

In this example, the principal difference between the two methods is the required minimum gradient, which is 10 percent less by the minimum tractive tension design method than by that based on minimum self-cleansing velocity.

9.7.2 Example 2

Use the simplified design approach outlined in subsection 9.3.6 to design the sewer in Example 1, but only for a population of 6000. Take the average household size (p) to be five.

The number of households is 6000/5, i.e. 1200. From Table 9.3 choose a 225 mm diameter sewer.

From equation (9.51) determine the peak flow:

$$\begin{aligned} q &= 0.009N \\ &= 0.009 \times 1200 \\ &= 10.8 \text{ l/s} \end{aligned}$$

Calculate the minimum sewer gradient from equation (9.47):

$$\begin{aligned} I_{\text{min}} &= 0.00518q^{-6/13} \\ &= 0.00518(10.8)^{-6/13} \\ &= 0.00173 \text{ (1 in 579)} \end{aligned}$$

9.7.3 Example 3

Design a sewer to take the sewage from 50 households with an average size of 20. The water supply is currently from yard taps and the consumption is 50 litres per caput per day. Allowance is to be made for a future upgrading of the water supply to multiple-tap, in-house connections, when the consumption is anticipated to rise to 120 litres per caput per day. Assume a return factor of 0.85, a peak wastewater flow factor of 1.8, and no infiltration.

Water consumption is expected to increase by 140 percent. From inspection of Table 9.1, choose:

initial $d/D = 0.44$, for which $q/Q = 0.4002$

final $d/D = 0.80$, for which $q/Q = 0.9773$

This permits an increase in water consumption of:

$$(0.9773 - 0.4002)/0.4002$$

i.e. 144 percent, which is satisfactory.

For $d/D = 0.44$, $k_a = 0.3328$ and $k_r = 0.2295$ (Table 9.2), so equations (9.39) and (9.31) can be solved as follows (to produce equations analogous to equations (9.47) and (9.48)), with q in l/s:

$$I_{\min} = 5.15 \times 10^{-3} q^{-6/13}$$
$$D = 0.00321(q/I_{\min}^{0.5})^{3/8}$$

The resulting equation analogous to equation (9.5) (with D now in millimetres), is:

$$q = 6.4 \times 10^{-5} D^{13/6}$$

The peak flow per household is given by equation (9.21) as:

$$1.8 \times 10^{-5} \times 20 \times 50$$

i.e. 0.018 l/s per household. Thus:

$$q = 0.018N$$

and

$$N = 3.56 \times 10^{-3} D^{13/6}$$

so that the following table (equivalent to Table 9.3) can be developed:

D(mm)	N
100	76
150	184
225	444
300	829

Thus, for 50 households, choose a sewer of 100 mm diameter. The peak flow is (0.018 × 50), i.e. 0.9 l/s. As this is < 2.2 l/s, take $q = 2.2$ l/s and thus the value of $I_{\min}$ is given by:

$$I_{\min} = 5.15 \times 10^{-3} q^{-6/13}$$
$$= 5.15 \times 10^{-3} (2.2)^{-6/13}$$
$$= 0.0036 \text{ (1 in 279)}$$

10

Sewage Treatment

10.1 INTRODUCTION

The sewage collected in settled sewerage and simplified sewerage networks (Chapters 8 and 9) should be treated prior to discharge to a surface watercourse (stream or river), or prior to reuse in either agriculture (crop irrigation) or aquaculture (fish farming). It may, of course, be more convenient to discharge this sewage into a conventional sewerage network if there is one nearby; if not, then it will have to be treated on its own.

Sewage treatment is a large subject in its own right, and it is not the purpose of this chapter to discuss all aspects of the subject. Instead, a fairly *brief* review of treatment in waste stabilization ponds and effluent storage reservoirs, and effluent reuse in agriculture and aquaculture, is given. The interested reader should refer to section 10.6, which gives a list of recommended further reading.

10.2 WASTE STABILIZATION PONDS

Waste stabilization ponds (Figure 10.1) are large shallow basins enclosed by earthen embankments in which raw sewage is treated by entirely natural processes involving both algae and bacteria. Since these processes are unaided by man (who merely allocates a place for their occurrence) the rate of oxidation is rather slow and, as a result, long hydraulic retention times are employed, 10–50 days not being uncommon. Ponds have considerable advantages—particularly regarding costs and maintenance requirements and the removal of faecal bacteria—over all other methods of sewage treatment. They are, without doubt, the most important method of sewage treatment in developing countries wherever sufficient land is

Figure 10.1 A 0.8 ha facultative waste stabilization pond at Thika, Kenya

available and where the temperature is most favourable for their operation.

There are three main types of waste stabilization ponds: anaerobic, facultative and maturation, and these are usually arranged in series (Figure 10.2). Anaerobic and facultative ponds are designed for the removal of organic matter (expressed in terms of its biochemical oxygen demand, BOD), and maturation ponds for the removal of excreted pathogens (although some BOD removal occurs in maturation ponds, and some pathogen removal in anaerobic and facultative

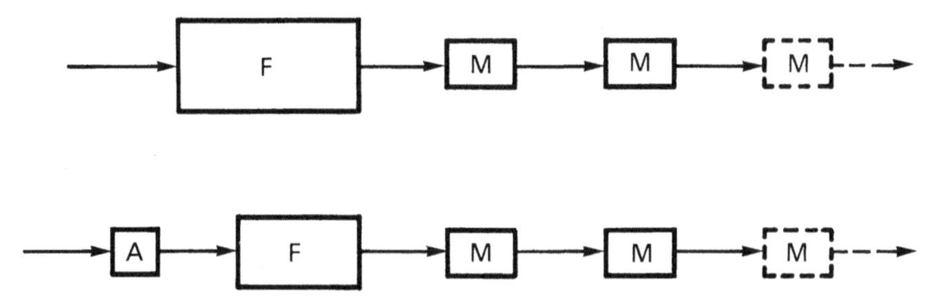

Figure 10.2 Waste stabilization ponds in series: a series normally comprises an anaerobic (*A*) pond followed by a facultative (*F*) and one or more maturation ponds (*M*). A pond complex may have several such series in parallel

ponds). Both facultative and maturation ponds are photosynthetic ponds, i.e. they contain high numbers of green algae, which produce the oxygen needed by the pond bacteria from their photosynthetic activities; the generalized 'equations' are:

$$CO_2 + H_2O \xrightarrow[\text{algae}]{\text{sunlight}} \text{carbohydrate} + O_2$$
$$BOD + O_2 \xrightarrow{\text{bacteria}} \text{oxidized waste} + CO_2$$

The process design of anaerobic, facultative and maturation ponds is outlined in subsections 10.2.1–10.2.3, with special reference to producing an effluent suitable for crop irrigation (section 10.3) or fish culture (section 10.5). Two of the books listed in section 10.6 provide advice on how ponds should be constructed and maintained.

10.2.1 Anaerobic ponds

An anaerobic pond is the first pond in a series of ponds. Anaerobic ponds are *extremely* efficient at removing BOD. As their name implies they are devoid of any dissolved oxygen, and they function much like open septic tanks (Chapter 6). Provided they are designed properly, they do *not* cause any odour nuisance. Design is on the basis of a permissible BOD loading rate and, for anaerobic ponds, it is the volumetric loading that is important. This varies with temperature, as does their performance expressed as percentage BOD removal (Table 10.1). Volumetric loading (λ_v) is expressed in g per m^3 per day, and is therefore given by:

$$\lambda_v = L_i Q / V_a \qquad (10.1)$$

where L_i = BOD of the raw sewage, mg/l (= g/m^3)
Q = flow rate, m^3/day
V_a = anaerobic pond volume, m^3

So, knowing the design temperature (see below) and hence the permissible loading rate from Table 10.1, the pond volume is readily determined. Anaerobic ponds are usually 2–5 m deep, so the pond area can be calculated, as can its effluent BOD (using the percentage removal given in Table 10.1), which is needed to design the facultative pond, which is the next pond in the series.

Table 10.1 Design values for the volumetric BOD loading rate on, and percentage BOD removal in, anerobic ponds at various temperatures

Temperature (°C)	Volumetric loading rate (g BOD/m³ day)	Percentage BOD removal
⩽ 10	100	40
11	120	42
12	140	44
13	160	46
14	180	48
15	200	50
16	220	52
17	240	54
18	260	56
19	280	58
20	300	60
21	300	62
22	300	64
23	300	66
24	300	68
⩾ 25	300	70

Design temperature

This is usually the mean temperature of the coldest month (or coldest quarter). (The mean temperature of any month is the mean of the mean daily temperatures, which are the means of the daily maximum and the daily minimum temperatures.) This is always the design temperature for the anaerobic and facultative ponds and, in the case of aquacultural reuse of the pond effluent, the fishponds; but, for agricultural reuse, the design temperature for the maturation ponds is usually the mean temperature of the coolest month in the irrigation season.

10.2.2 Facultative ponds

Facultative ponds are dark green in colour, and this is due to the micro-algae that grow naturally and profusely in it. The algal concentration is usually expressed in terms of chlorophyll *a*, the most important algal pigment used for photosynthesis; its concentration in facultative ponds is usually 500–1500 μg/l. The algae, through their photosynthetic activities, are responsible for most of the dissolved oxygen in the pond (some comes from the atmosphere through the pond surface). This oxygen is used by the heterotrophic bacteria in the pond to remove BOD, and these bacteria release carbon dioxide, which is fixed by the algae as they photosynthesize. Thus, there is a relationship of mutual benefit between the pond algae and the pond bacteria (Figure 10.3). Because sunlight (the raw energy needed for algal photosynthesis) arrives at the pond surface, facultative ponds are designed on the basis of the surface BOD loading rate (rather than the volumetric loading used for anaerobic ponds in which there are no algae and hence no photosynthesis). The permissible BOD surface loading (λ_s) varies with temperature (Table 10.2) and, since it is normally expressed in terms of kg BOD per hectare per day, it is given by:

$$\lambda_s = L_i Q \times 10^{-3}\ (\text{kg/d})/A_f \times 10^{-4}\ (\text{ha})$$
$$= 10 L_i Q / A_f \qquad (10.2)$$

where L_i = effluent BOD from the anaerobic pond, mg/l (= g/m^3)
A_f = facultative pond area, m^2

Equation (10.2) and Table 10.2 permit the facultative pond area to be calculated. Facultative ponds are usually 1.5–1.8 m deep, and they achieve a BOD removal of around 50 percent irrespective of temperature. However, some 70–90 percent of

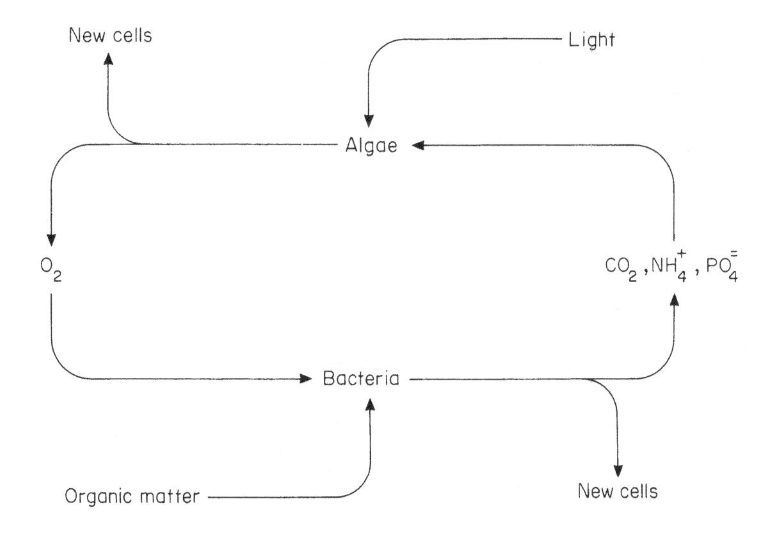

Figure 10.3 The mutualistic relationship between pond algae and heterotrophic bacteria: the algae produce oxygen, which is used by the bacteria, which produce carbon dioxide, which is used by the algae

Table 10.2 Design values for the surface BOD loading rate on facultative ponds at various temperatures

Temperatures (°C)	Surface loading rate (kg BOD/ha day)
$\leqslant$ 10	100
11	112
12	124
13	137
14	152
15	167
16	183
17	199
18	217
19	235
20	253
21	272
22	292
23	311
24	331
25	350
26	369
27	389
28	406
29	424

the BOD of the effluent from a facultative effluent is due to the algae present in it—around 3 mg BOD for every 100 μg chlorophyll *a* present. So the performance of a facultative pond is better expressed on the basis of *filtered* BOD (the effluent is filtered through a suitable glass-fibre filter paper, prior to BOD analysis, to remove the algae present): the removal of filtered BOD in a facultative pond is usually > 90 percent.

Helminth egg removal

At this stage in the design, it is sensible to check the removal of helminth eggs if the pond effluent is to be used for crop irrigation or fish culture (sections 10.3 and 10.5). The effluent requirements are $\not> 1$ intestinal nematode egg per litre for crop irrigation and 0 viable trematode eggs per litre for fish culture.

Table 10.3 gives the percentage removal of intestinal nematode eggs in a pond for various retention times. The retention time (θ, days) is simply the pond volume (V, m^3) divided by the sewage flow (Q, m^3/day). So for the anaerobic and facultative ponds, it is given by:

$$\theta_a = V_a/Q \qquad (10.3)$$

$$\theta_f = A_f D_f/Q \qquad (10.4)$$

where D_f = depth of facultative pond, m.

Table 10.3 is now used to determine the percentage egg removal first in the anaerobic pond (R_a) and then in the facultative pond (R_f). The number of eggs in the facultative pond effluent (E_f, per litre) is given by:

$$E_f = E_i[1 - (R_a/100)][1 - (R_f/100)] \qquad (10.5)$$

where E_i = number of eggs per litre of the raw sewage.

Values of E_i vary widely depending on the prevalence of intestinal nematode infection in the community: it can be as low as 10 per litre, or as high as 2000 per litre. A reasonable design value for a low-income urban community is around 1000 per litre. If equation (10.5) indicates that E_f is >1 per

Table 10.3 Values for the percentage removal (R) of intestinal nematode eggs in waste stabilization ponds at various retention times (θ, days)

θ	R	θ	R	θ	R
1.0	74.67	4.0	93.38	9.0	99.01
1.2	76.95	4.2	93.66	9.5	99.16
1.4	79.01	4.4	93.40		
1.6	80.87	4.6	94.85	10	99.29
1.8	82.55	4.8	95.25	10.5	99.39
2.0	84.08	5.0	95.62	11	99.48
2.2	85.46	5.5	96.42	12	99.61
2.4	87.72			13	99.70
2.6	87.85	6.0	97.06	14	99.77
2.8	88.89	6.5	97.57	15	99.82
3.0	89.82	7.0	97.99	16	99.86
3.2	90.68	7.5	98.32	17	99.88
3.4	91.45			18	99.90
3.6	92.16	8.0	98.60	19	99.92
3.8	92.80	8.5	98.82	20	99.93

litre, then one or more maturation ponds are required; if the maturation pond effluent is to be used for restricted irrigation (section 10.3), then the maturation pond is required only for egg removal, and a retention time of 3–5 days should be used for R_m in the equation:

$$E_m = E_f[1 - (R_m/100)]^n \qquad (10.6)$$

where E_m = number of eggs per litre of maturation pond effluent
R_m = percentage of egg removal in the maturation pond
n = number of maturation ponds (try $n = 1$ first, then $n = 2$, etc.)

If the pond effluent is to be used for fish culture, equations (10.5) and (10.6) can be used for trematode egg removal on the assumption that the values given in Table 10.3 for intestinal nematode egg removal can also be used for trematode egg removal (this is not unreasonable, despite there being very few data on the removal of trematode eggs in ponds).

10.2.3 Maturation ponds

Maturation ponds are designed primarily for faecal bacterial removal, using faecal coliform bacteria as the indicator organism for faecal bacterial pathogens (such as salmonellae, campylobacters—subsection 2.2.2). For both crop irrigation and fish culture, the maximum number of faecal coliforms is 1000 per 100 ml of pond effluent (for fish culture, per 100 ml of fishpond contents). The removal of faecal coliforms in ponds is best modelled on the assumption that it follows first-order kinetics in a completely mixed reactor; so, for a single pond:

$$N_e = N_i/(1 + k_T\theta) \qquad (10.7)$$

For a series of ponds comprising an anaerobic pond, a facultative pond and n maturation ponds, equation (10.7) becomes:

$$N_e = N_i/(1 + k_T\theta_a)(1 + k_T\theta_f)(1 + k_T\theta_m)^n \qquad (10.8)$$

where N_e = number of faecal coliforms per 100 ml of effluent
N_i = number of faecal coliforms per 100 ml of raw sewage (usually 10^7–10^8 per 100 ml)
k_T = first-order rate constant for faecal coliform removal, per day ('reciprocal days')
θ = pond retention time (the subscripts a, f and m refer to the anaerobic, facultative and maturation ponds, respectively)
n = number of maturation ponds

The value of k_T is highly temperature-dependent, and is given by:

$$k_T = 2.6\,(1.19)^{T-20} \tag{10.9}$$

Equation (10.9) gives the value of k_T at 20 °C as 2.6 per day, and shows that k_T varies by 19 percent for every change in temperature of 1 °C. Values for k_T for 10–30 °C are given in Table 10.4.

By the time the maturation ponds are to be designed, the anaerobic and facultative ponds will have already been designed, so θ_a and θ_f are known, and the only unknowns in equation (10.8) are θ_m and n. One equation and two unknowns means that there is no exact solution; instead, equation (10.8) is rewritten as:

$$\theta_m = \{[N_i/N_e(1 + k_T\theta a)(1 + k_T\theta_f)]^{1/n} - 1\}/k_T \tag{10.10}$$

Table 10.4 Values of the first-order rate constant for faecal coliform removal (k_T) at various temperatures

T (°C)	k_T (per day)	T (°C)	k_T (per day)
11	0.54	21	3.09
12	0.65	22	3.68
13	0.77	23	4.38
14	0.92	24	5.21
15	1.09	25	6.20
16	1.30	26	7.38
17	1.54	27	8.77
18	1.84	28	10.46
19	2.18	29	12.44
20	2.60	30	14.81

and solved first for $n = 1$ then for $n = 2$ and $n = 3$, etc. Values of $\theta_m > \theta_f$ are ignored, as are values < 3 days, which is the minimum acceptable value to permit good algal growth and minimize hydraulic shortcircuiting. Also, the BOD loading on the first maturation pond must not be greater than 75 percent of that on the facultative pond; therefore:

$$\lambda_{s(m1)} = 10\ L_i Q/A_{m1} \qquad (10.11)$$

where $\lambda_{s(m1)}$ = BOD loading on first maturation pond, kg/ha day (= $\frac{3}{4}$ value given in Table 10.2)
L_i = BOD of effluent from facultative pond (based on Table 10.1 and 50 percent removal in the facultative pond)
A_{m1} = area of first maturation pond, m^2

Since $\theta = V/Q$, i.e. AD/Q, $Q/A = D/\theta$ and equation (10.11) can be rewritten as follows, to give the minimum value of the retention time in the first maturation pond:

$$\theta_{m1} = 10\ L_i D_m/\lambda_{s(m1)} \qquad (10.12)$$

If the pond effluent is not to be used for either crop irrigation or fish culture, but is to be discharged into a surface water (stream, river, lake), then N_e does not have to be specified as strictly as 1000 per 100 ml, and the BOD of the final effluent can be calculated on the basis of 25 percent removal in each of the maturation ponds.

The design example given in section 10.7 shows how these equations are used in practice.

10.3 AGRICULTURAL REUSE

Sewage is being increasingly used, particularly in arid and semi-arid areas, for crop irrigation: sewage contains not only water but also crop nutrients (especially nitrogen and phosphorus), so crop yields are higher (Table 10.5). However, if untreated sewage is used to irrigate crops, there are real risks to public health—the health of the fieldworkers, and the health of the crop consumers. The World Health Organization

Table 10.5 Crop yields obtained in India after irrigation with waste stabilization pond effluent and freshwater supplemented with NPK fertilizer

Irrigation water	Crop yield (tonnes/ha year)				
	Wheat	Beans	Rice	Potato	Cotton
Freshwater +NPK	2.70	0.78	2.03	17.16	1.70
Waste stabilization pond effluent	3.45	0.72	2.98	22.31	2.41

has set guidelines for the microbiological quality of *treated* sewage to be used for crop irrigation, as follows:

(a) for restricted irrigation (i.e. of crops not for direct human consumption):

$\ngtr$ 1 intestinal nematode egg per litre

(the intestinal nematodes are the geohelminths: *Ascaris, Trichuris* and the human hookworms—see subsection 2.2.2.)

(b) for unrestricted irrigation (i.e. including crops eaten uncooked):

$\ngtr$ 1 intestinal nematode egg per litre, and

$\ngtr$ 1000 faecal coliform bacteria per 100 ml

The nematode guideline is to protect the health of both fieldworkers and consumers, and the faecal coliform guideline to protect the health of consumers (fieldworkers are not, in general, at risk from faecal bacterial infections).

Treatment in waste stabilization ponds and effluent storage reservoirs (sections 10.2 and 10.4) readily achieves these guideline qualities. The method of application of the treated sewage to the crop is also important: spray irrigation should not be used, generally only furrow irrigation. If localized (i.e. drip, trickle) irrigation is practised, then the guideline values are not so applicable: treated sewage is applied only around the root zone of the crop (Figure 10.4), so the health risk to the fieldworkers and consumers is minimal. Localized irrigation is highly economic with irrigation water as very little is

Figure 10.4 Localized irrigation of cotton: note the discharge of treated wastewater from the emitter around the root zone of the plant

lost by evaporation, so a greater area can be irrigated with the same amount of water, but it costs more. Nonetheless, it is widely used in some water-short countries, such as Israel where over 80 percent of all sewage is used for crop irrigation (mainly of cotton as finished cotton products are a major export earner for Israel).

10.4 EFFLUENT STORAGE RESERVOIRS

The effluent from waste stabilization ponds can only be used for crop irrigation during the irrigation season; during other months it has to be discharged to a surface water. Effluent storage reservoirs enable the whole year's effluent to be used for irrigation, so that a larger area can be irrigated and more crops produced (Figure 10.5).

For restricted irrigation, the sewage is first treated in an anaerobic pond (subsection 10.2.1) and then discharged into a single effluent storage reservoir. The reservoir is full at the

Figure 10.5 Effluent storage reservoir in Israel, showing effluent take-off pumps

beginning of the irrigation season and empty at the end of it. So the reservoir volume (V_{esr}, m^3) is determined on the basis of storing the anaerobic pond effluent during the non-irrigation season:

$$V_{esr} = 30\,(12 - m)Q \tag{10.13}$$

where m = number of months in the irrigation season
Q = sewage flow, m^3/day

Whilst the effluent is stored in the reservoir, algae grow and the reservoir contents are much like a facultative or maturation pond. All intestinal nematode eggs settle out in the anaerobic pond and in the reservoir, so that the effluent meets the WHO guideline value of $\ngtr 1$ egg per litre.

For unrestricted irrigation, a single reservoir is insufficient since, although its contents may have < 1000 faecal coliforms per 100 ml at the start of the irrigation season, its effluent during the irrigation will have more as it is continually receiving a high faecal coliform input from the anaerobic pond. Three or four storage reservoirs in parallel are needed, depending on the length of the irrigation season, and these are

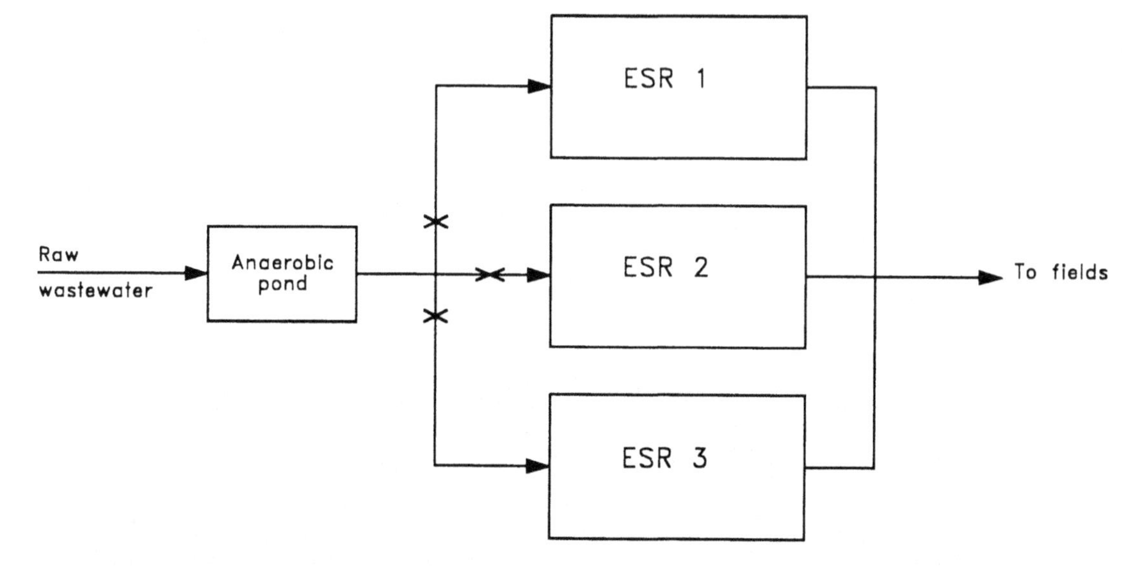

Figure 10.6 Sequential batch-fed effluent storage reservoirs for unrestricted irrigation. The wastewater is pretreated in an anaerobic pond and then discharged into one of three (or four) effluent storage reservoirs in parallel, each of which is operated on a cycle of fill–rest–use in order to produce an effluent that meets the WHO guideline quality of < 1000 faecal coliforms per 100 ml

operated in a sequential batch-fed mode (Figure 10.6). Each reservoir operates on a cycle of fill–rest–use (Table 10.6), and faecal coliform die-off to < 1000 per 100 ml occurs during the fill and rest periods of the cycle, so that during the use period its effluent complies with the guideline.

10.5 AQUACULTURAL REUSE

Certain species of fish, mainly carp and tilapia, grow extremely well in sewage-fertilized fishponds. Yields of over 10 tonnes of fish per ha of fishpond area per year can easily be obtained from well-managed fishponds; even ponds not-so-well-managed can produce two to six tonnes of fish per ha per year.

The World Health Organization has given the following guidelines for the microbiological quality of effluents used to fertilize fishponds:

viable trematode eggs: zero per litre

faecal coliform bacteria: 1000 per 100 ml of fishpond water

Table 10.6 Operation of three effluent storage reservoirs in parallel for an irrigation season of six months (April–September)

Month	ESR 1	ESR 2	ESR 3
January	Rest	Fill (1)	Empty
February	Rest	Rest	Fill (1)
March	Rest	Rest	Fill (1)
April	**Use**	Rest	Fill (1)
May	**Use**	Rest	Fill (1)
June	Fill (1)[a]	**Use**	Rest
July	Fill (1)	**Use**	Rest
August	Fill ($\frac{1}{2}$)	Fill ($\frac{1}{2}$)	**Use**
September	Fill ($\frac{1}{2}$)	Fill ($\frac{1}{2}$)	**Use**
October	Fill ($\frac{1}{3}$)	Fill ($\frac{1}{2}$)	Empty
November	Fill ($\frac{1}{2}$)	Fill ($\frac{1}{2}$)	Empty
December	Rest	Fill (1)	Empty
Months storage	4	4	4

[a]Proportion of monthly flow diverted to each ESR

Note: (1) Here the hot season is assumed to be June and July, so ESR 3 has a two months' rest period in these months, the others have four in the cooler parts of the year. (2) Alternative management strategies are possible, depending on what volume is required at which part of the irrigation season. In the table, equal volumes are assumed to be needed in each two-month part of the irrigation season (i.e. each reservoir has a four-month storage volume). But this could be readily altered to, say, 3-, 4- and 5-month storage volumes if so required by the pattern of crop water demand (ESR 1 would also rest in November and December, and ESR 3 would receive half the flow in November and December but none in January or February).

The trematode worms of importance in sewage-fed aquaculture are (see subsection 2.2.2):

Schistosoma mansoni, S. haematobium and *S. japonicum*

Clonorchis sinensis, and

Fasciolopsis buski

To achieve this microbiological quality, the sewage should be treated in an anaerobic and a facultative pond prior to discharge into a fishpond (the anaerobic pond is omitted for sewage collected in a settled sewerage network). These ponds are designed as in subsections 10.2.1 and 10.2.2. The fishpond is then designed on the basis of a total nitrogen loading of 4 kg N per ha per day (if the nitrogen loading is too high, the algal biomass concentration is correspondingly too high and there is the consequent danger of zero dissolved oxygen in the fishpond at night, and thus there may be fish kills; if the nitrogen loading is too low, the algal biomass is then too low as well and the fish yield is also low: a nitrogen loading of 4 kg N per ha per day is about optimal). To determine the total nitrogen loading on the fishpond, the total nitrogen concentration in the facultative pond effluent (TN_e, mg/l) must be known; this is given by Reed's equation:

$$TN_e = TN_i \exp\{-[0.0064(1.039)^{T-20}][\theta_f + 60.6(\mathrm{pH} - 6.6)]\} \quad (10.14)$$

where TN_i = total nitrogen concentration in the influent to the facultative pond, mg/l (since there is essentially no removal of total nitrogen in the anaerobic pond, TN_i = total nitrogen concentration in the raw sewage)

The facultative pond pH is estimated from the equation:

$$\mathrm{pH} = 7.3 \exp(0.0005A) \quad (10.15)$$

The fishpond area is then calculated from equation (10.2), using TN_e instead of L_i and with λ_s = 4 kg N per ha per day; and then the fishpond retention time (θ_{fp}, days) is calculated from equation (10.4) using a depth of 1 m.

The faecal coliform concentration in the fishpond (N_{fp}, per 100 ml) is now determined from the following form of equation (10.8):

$$N_{fp} = N_i/(1 + k_T\theta_a)(1 + k_T\theta_f)(1 + k_T\theta_{fp}) \quad (10.16)$$

If $N_{fp} > 1000$ (the WHO guideline value for aquacultural reuse), θ_{fp} is increased until $N_{fp} = 1000$. The design example given in section 10.7 shows how these equations are used.

10.5.1 Fish yields

Fish (tilapia or carp) are stocked at the rate of three fingerlings (each weighing around 20 g) per m^2 of fishpond area. After three months these will have grown to 150–250 g, the size most commonly bought and consumed by low-income communities (Figure 10.7). The pond is then drained, the fish harvested and the pond refilled and restocked.

Assuming that there are three harvests per year, and that

Figure 10.7 Indian major carp being harvested from a sewage-fed fishpond in Calcutta, India. Some 13 000 tonnes of fish are harvested per year from these ponds, which cover an area of 3000 ha and use 550 000 m^3 of sewage per day; this represents around 17 percent of the local demand for fish

there is a 25 percent fish loss due to mortality, poaching and consumption by fish-eating birds, the fish yield would be:

$$3 \text{ harvests/year} \times 0.75 \text{ (efficiency)} \times (3 \times 200 \text{ g fish/m}^2) \times 10^{-3} \text{ kg/g} \times 10^4 \text{ m}^2\text{/ha} = 13\,500 \text{ kg fish per ha per year}$$

In India, the species of Indian major carp grown in sewage-fed fishponds (usually catla, *Catla catla*; mrigal, *Cirrhina mrigala*; and rohu, *Labeo rohita*) are sold by the fish farmers for around US$0.54 per kg, so the aquacultural resuse of pond effluent can be quite a profitable business. It also increases peri-urban food production and creates employment.

10.6 FURTHER READING

R. M. Ayres and D. D. Mara, *Analysis of Wastewater for Use in Agriculture: A Laboratory Manual of Parasitological and Bacteriological Techniques*. World Health Organization, Geneva (1995).

P. Edwards, *Reuse of Human Wastes in Aquaculture: A Technical Review*. Water and Sanitation Report No. 2. The World Bank, Washington, DC (1992).

P. Edwards and P. S. V. Pullin, *Wastewater-fed Aquaculture*. Environmental Sanitation Information Center, Asian Institute of Technology, Bangkok (1990).

Health Guidelines for the Use of Wastewater in Agriculture and Aquaculture. Technical Report Series No. 778. World Health Organization, Geneva (1989) (also available in French and Spanish).

N. Khouri, J. M. Kalbermatten and C. R. Bartone, *Reuse of Wastewater in Agriculture: A Guide for Planners*. Water and Sanitation Report No. 6. The World Bank, Washington , DC (1994).

D. D. Mara, *Sewage Treatment in Hot Climates*. John Wiley & Sons, Chichester (1976; 2nd edition forthcoming, 1997).*

D. D. Mara and A. M. Cairncross, *Guidelines for the Safe Use of Wastewater and Excreta in Agriculture and Aquaculture: Measures for Public Health Protection*. World Health Organisation, Geneva (1989) (also available in French and Spanish).

D. D. Mara and S. W. Mills, "Who's afraid of anaerobic ponds?" *Water Quality International* (27), 34–36 (1994).

D. D. Mara and H. W. Pearson, "Sequential batch-fed effluent storage reservoirs: a novel method of wastewater treatment prior to unrestricted irrigation". *Water Science and Technology* **26** (7/8), 1459–1464. (1992).

D. D. Mara, G. P. Alabaster, H. W. Pearson and S. W. Mills, *Waste Stabilization Ponds: A Design Manual for Eastern Africa*. Lagoon Technology International, Leeds (1992).*

D. D. Mara, P. Edwards, D. Clark and S. W. Mills, "A rational approach to the design of wastewater-fed fishponds". *Water Research* **27**, 1797–1799 (1993).
G. S. Pettygrove and T. Asano, *Irrigation with Reclaimed Municipal Wastewater: A Guidance Manual*. Lewis Publishers Inc., Chelsea, MI (1985).
H. I. Shuval, A. Adin, B. Fattal, E. Rawitz and P. Yekutiel, *Wastewater Irrigation in Developing Countries: Health Effects and Technical Solutions*. Technical Paper No. 51. The World Bank, Washington, D.C. (1986).

*These two books provide advice on pond construction and O&M.

10.7 DESIGN EXAMPLE

Design a waste stabilization pond to treat sewage from a low-income settlement of 5000 people. The BOD contribution per person per day is 40 g, and the sewage flow 80 litres per person per day. The mean temperature of the coolest month of the year is 20 °C and during the irrigation season 25 °C. Assume that the faecal coliform and intestinal nematode egg counts in the raw sewage are 1×10^8 per 100 ml and 750 per litre, respectively, and that the final effluent is to be used for (a) restricted crop irrigation, (b) unrestricted crop irrigation, and (c) fish culture. For the latter, take the total nitrogen concentration and the alkalinity of the raw sewage as 60 mg N/l and 300 mg $CaCO_3$/l, respectively.

10.7.1 Solutions

The BOD (L_i) is $(40\ \text{g} \times 10^3)\ \text{mg}/80\ \text{l} = 500\ \text{mg/l}$ and the sewage flow (Q) is $5000 \times 80 \times 10^{-3} = 400\ \text{m}^3/\text{d}$.

Anaerobic pond

For 20 °C the volumetric BOD loading rate is given in Table 10.1 as 300 g/m^3 day. The anaerobic pond volume is given by equation (10.1) as:

$$\begin{aligned} V_a &= L_i Q/\lambda_v \\ &= 500 \times 400/300 = 667\ \text{m}^3 \end{aligned}$$

So, assuming a depth of 2 m, the area is 333 m^2. From Table 10.1 the BOD removal is 60 percent, so the BOD of the anaerobic pond effluent is (0.4 × 500), i.e. 200 mg/l. The retention time in the anaerobic pond is given by equation (10.3) as:

$$\theta_a = V_a/Q$$
$$= 667/400 = 1.7 \text{ days}$$

Facultative pond

The BOD surface loading for 20 °C is given in Table 10.2 as 253 kg/ha day. The facultative pond area is given by equation (10.2) as:

$$A_f = 10\ L_i Q/\lambda_s$$
$$= 10 \times 200 \times 400/253 = 3162 \text{ m}^2$$

and the retention time is given by equation (10.4) assuming a depth of 1.5 m, as:

$$\theta_f = A_f D_f/Q$$
$$= 3162 \times 1.5/400 = 11.9 \text{ days}$$

Intestinal nematode egg removal

From Table 10.3, the egg removal in the anaerobic pond ($\theta_a = 1.7$ days) is 81.7 percent, and in the facultative pond ($\theta_f = 11.9$ days) 99.6 percent. So the number of eggs per litre of the facultative pond effluent is given by equation (10.5) as:

$$E_f = E_i[1 - (R_a/100)][1 - (R_f/100)]$$
$$= 750(1 - 0.817)(1 - 0.996) = 0.5/l$$

Since this is <1 egg per litre, the WHO guideline for restricted irrigation is achieved by the anaerobic and facultative pond, and therefore no maturation pond is required. However, for unrestricted irrigation, maturation ponds will be required to reduce the faecal coliform level to 1000 per 100 ml.

Maturation ponds

These are designed for 25 °C (the mean temperature of the coolest month in the irrigation season). From equation (10.9) the value of k_T at 25 °C is given by:

$$k_T = 2.6\ (1.19)^{T-20}$$
$$= 2.6\ (1.19)^5 = 6.2 \text{ per day}$$

The minimum value of the retention time in the first maturation pond is given by equation (10.12) (assuming 60 percent BOD removal in the anaerobic pond and 50 percent in the facultative pond, and a maturation pond depth of 1 m) as:

$$\theta_{m1} = 10 L_i D_m / 0.75 \lambda_{s(fac)}$$
$$= 10 \times 0.4 \times 0.5 \times 500 \times 1/0.75 \times 350$$
$$= 3.8 \text{ days}$$

Note that the value used here for $\lambda_{s(fac)}$ is 350 kg/ha day which is the value for 25 °C (Table 10.2).

Using equation (10.10):

$$\theta_m = \{[N_i/N_e(1 + k_T\theta_a)\ (1 + k_T\theta_f)]^{1/n} - 1\}/k_T$$
$$= \{[10^8/1000(1 + 6.2 \times 1.7)(1 + 6.2 \times 11.9)]^{1/n} - 1\}/6.2$$

For: $n = 1$, $\theta_m = 18.5$ days (ignore as $> \theta_f$)
$n = 2$, $\theta_m = 1.6$ days (ignore as < 3 days)

So choose $\theta_{m1} = 3.8$ days and $\theta_{m2} = 3$ days, these being the minimum values for the first maturation pond and for subsequent ones, respectively. Use equation (10.8) to check the resulting value of N_e:

$$N_e = N_i/(1 + k_T\theta_a)(1 + k_T\theta_f)\ (1 + k_T\theta_{m1})(1 + k_T\theta_{m2})$$
$$= 10^8/(1 + 6.2 \times 1.7)(1 + 6.2 \times 11.9)(1 + 6.2 \times 3.8)$$
$$\times (1 + 6.2 \times 3)$$
$$= 241 \text{ per 100 ml (satisfactory)}$$

So, for unrestricted irrigation, two maturation ponds are required, with retention times of 3.8 and 3.0 days. The corresponding areas are given by the following version of equation (10.4):

$$A_m = Q\theta_m/D_m$$

$$A_{m1} = 400 \times 3.8/1 = 1520 \text{ m}^2$$

$$A_{m2} = 400 \times 3.0/1 = 1200 \text{ m}^2$$

Fish culture

Use the anaerobic and facultative ponds as calculated above. Calculate the total nitrogen concentration in the facultative pond effluent from equations (10.15) and (10.14):

$$\text{pH} = 7.3 \exp(0.0005A)$$

$$= 7.3 \exp(0.0005 \times 300) = 8.5$$

$$TN_e = TN_i \exp\{-[0.0064(1.039)^{T-20})][\theta_f + 60.6(\text{pH} - 6.6)]\}$$

$$= 60 \exp\{-[0.0064(1.039)^5][11.9 + 60.6(8.5 - 6.6)]\}$$

$$= 22 \text{ mg/l}$$

So the fishpond area is given by equation (10.2), with $\lambda_s = 4$ kg N/ha day, as

$$A_{fp} = 10(TN_e)Q/\lambda_s$$

$$= 10 \times 22 \times 400/4 = 22\,000 \text{ m}^2$$

The retention time in the fishpond is given by the following version of equation (10.4), taking the depth as 1 m:

$$\theta_{fp} = A_{fp}D_{fp}/Q$$

$$= 22\,000 \times 1/400 = 55 \text{ days}$$

Check the faecal coliform concentration in the fishpond from equation (10.16):

$$N_{fp} = N_i(1 + k_T\theta_a)(1 + k_T\theta_f)(1 + k_T\theta_{fp})$$
$$= 10^8/(1 + 6.2 \times 1.7)(1 + 6.2 \times 11.9)(1 + 6.2 \times 55)$$
$$= 339 \text{ per 100 ml (satisfactory as} < 1000)$$

10.7.2 Summary

(a) *For restricted irrigation:*

anaerobic pond	667 m^2	1.7 days
facultative pond	3162 m^2	11.9 days
Total	3829 m^2	13.6 days

(b) *For unrestricted irrigation:*

anaerobic pond	667 m^2	1.7 days
facultative pond	3162 m^2	11.9 days
1st maturation pond	1520 m^2	3.8 days
2nd maturation pond	1200 m^2	3.0 days
Total	6549 m^2	20.4 days

(c) *For fish culture:*

anaerobic pond	667 m^2	1.7 days
facultative pond	3162 m^2	11.9 days
fishpond	22 000 m^2	55.0 days
Total	25 829 m^2	68.6 days

11

Sociocultural Aspects

11.1 INTRODUCTION

Using a sanitation facility is a private matter in all societies, but different societies may have different preferences and practices. Much of what has been written about the sociocultural aspects of sanitation—what one might term 'ethnocoprology'—is about *rural* sanitation. In urban areas the situation is rather easier: the much higher population density and the much greater lack of alternative defecation (and even urination) sites mean that perceived needs for sanitation are correspondingly greater in urban areas than in rural areas. People clearly prefer a faecally uncontaminated environment, and in urban areas without adequate sanitation the faecal density soon rises to socially unacceptable levels. However, the 'germ theory' of disease may not be understood in all urban societies, especially by newly arrived immigrants from rural areas, and thus the real dangers to health posed by human (and, to a lesser extent, animal) excreta may not be fully appreciated. Even so, the 'dirtiness' and odour of faeces are apparent to all, and urban people generally need little motivation and encouragement to have their own sanitation facility, provided, of course, that it is affordable and socioculturally what they want. Indeed, having a latrine or toilet may often be perceived as conferring status on the household. Chapter 12 examines the socioeconomic and financial aspects of sanitation, and the remaining sections in this chapter look at the common sociocultural constraints to be addressed in urban sanitation programmes.

11.2 PRIVACY

Everyone prefers to use a latrine or toilet in private, so the system *must* ensure privacy. If the latrine has a door, then it should extend right down to the floor, as in some societies people do not like their feet to be seen from outside. People—especially women—may also feel that their privacy is compromised if they are seen obviously going to the latrine—for example, when they are carrying water or leaves for anal cleansing.

11.3 POSTURE, CLEANSING AND ORIENTATION

Some people prefer to sit whilst defecating, others prefer to squat. Whilst one can make a good physiological argument in favour of squatting (better evacuation of faeces), life is too short to persuade 'sitters' to become 'squatters'. Those who sit should simply be provided with pedestal seat units, and those who squat with squat-pans. What should not happen is that squatters are given pedestal seat units (Figure 11.1). Women may not like to squat if they find that squatting dirties their clothes.

Similarly, some people use water for anal cleansing but others prefer to use paper (or other absorbent material). Again, 'washers' should be accommodated for, as should 'wipers'. The former need a convenient source of water; this is not usually a problem in households with an on-plot water supply but, for those relying on a hand-carried water supply (from a public standpipe or well, for example), arrangements should be made for a suitably sized water storage vessel to be located in or near the latrine (this is a point that needs to be discussed with the users: see Chapter 14). Wipers need access to a supply of cheap toilet paper, or their latrine must be compatible with alternative anal cleansing materials; bulky materials, such as corncobs, are not compatible with pour-flush toilets, and an easily cleaned receptacle must be provided in this case and the faecally contaminated material disposed of hygienically. Again, this must be discussed with the users (see Chapter 14).

Orientation of the body during defecation may also be important. For example, Muslims generally do not defecate in a position such that they are facing, or have their backside towards, Mecca. This should be discussed with the local imam or mullah, and his advice followed.

Figure 11.1 Those who prefer to squat during defecation should not be provided with pedestal seat units!

11.4 GENDER AND AGE

It is important to identify any gender constraints. With household-level latrines these are not commonly a problem but with compound-level latrines they may be, especially when several unrelated families live in the same compound. In some societies some female members of the family may not use the same latrine as the men: for example, a daughter entering puberty (i.e. at menarche) may not be allowed—or may prefer not—to use the same latrine as her father; or a husband may not allow his mother-in-law to use the same latrine that he uses.

Another gender aspect of importance is to ascertain what women use to absorb menstrual blood and whether this is disposed of in the latrine; if it is, then the system must be able to cope with it. This is especially important in the case of simplified sewerage as the sewer gradients are often quite flat (see Chapter 9).

Age can also be important: are small children, for example, encouraged to use or discouraged from using the same facility as older members of the family? Often they are discouraged as it is felt that they will make it dirty (Figure 11.2). The answer here is that children should be properly toilet-trained so that they do not dirty the toilet—children should always use the toilet or latrine as they are more susceptible to excreted infections and their faeces are often highly charged with excreted pathogens. If not properly disposed of in the toilet or latrine, their faeces will pose a real health risk to all other members of the family and any visitors.

Figure 11.2 Children are often discouraged from using a latrine or toilet, but this is *not* wise: their excreta are often highly charged with excreted pathogens and, if not disposed of in the latrine, will pose a real health risk to all other family members

11.5 MAINTENANCE

HLOM (see section 1.2) is not usually a problem (but see Chapter 14). However, with compound-level sanitation facilities it can be, especially when several unrelated families live in the same compound. The problem here is mainly fouling of the system by young children, and women not wanting to clean up after other women's children. This is clearly another point that must be discussed with the users (see Chapter 14).

11.6 FURTHER READING

L. Box, A. Moussane, E. Sizoo and N. Vink, *Culture and Communication: The Forgotten Dimension in Development Co-operation*. Bulletin No. 329, Royal Tropical Institute, Amsterdam (1993).

H. Perrett, *Social Feasibility Analysis of Low-cost Sanitation Projects*. TAG Technical Note No. 5. The World Bank, Washington, DC (1983).

H. Perrett, *Involving Women in Sanitation Projects*. TAG Discussion Paper No. 3. The World Bank, Washington, DC (1985).

M. Simpson-Hebert, *Methods for Gathering Socio-cultural Data for Water Supply and Sanitation Projects*. TAG Technical Note No. 2. The World Bank, Washington, DC (1983).

S. Caincross and V. Kochar, *Studying Hygiene Behaviour: Methods, Issues and Experiences*. Sage Publications, New Delhi, (1994).

12

Socioeconomic Aspects

12.1 ECONOMIC COSTING

Ideally, a cost-benefit analysis should be used to compare different sanitation technologies, but it is virtually impossible to quantify all the benefits (such as improved health, user convenience). The only alternative is to determine the cost of each technology and select the one with the least cost unless the users are willing to pay for one with a higher cost. Clearly, costs have to be discussed with the community, as well as all the details of the technology options, so that they can decide what they want and what they will pay (see Chapter 15).

However, there is a difficulty here: the costs that the users pay is a financial cost, and this can be artificially low. A good example of this is water: the cost of water to consumers reflects the historical mix of investment costs made by the water supply authority, usually over several decades—a town or city exploits its cheapest source of water first, and then over time the development of new water sources becomes progressively more expensive. For the water authority, and especially the national government, it is the full cost of the most recently exploited cubic meter of water that is important; this is what economists call the marginal cost of water, and this is higher than the cost of water to the consumer.

With sanitation technologies (and indeed any other public sector investment whose benefits are not fully quantifiable) what is needed is a method of determining their real costs—not (at least in the first instance) the cost that the users would pay, but the cost of each alternative sanitation technology under consideration to the national economy, i.e. to the natural resource endowment of the country. This is a very important consideration: for example, local engineers might favour conventional sewerage, but its dependence on large volumes of flushing water might place too great a demand on local (even national) water resources and mean that too much of the country's money would have to be spent on exploiting those water resources in order to provide sufficient water for flushing, and the Ministry of Economic Development (or whatever it might be called) might have to step in and say that the country—the national economy—cannot afford it as its economic cost is too high.

So, what is done first with competing sanitation alternatives is to evaluate their *economic* costs and then determine what the users will pay (these are *financial* costs—see section 12.2). The purpose of this economic costing is to give policy-makers a proper economic basis for their decisions by providing a 'price tag' for a given sanitation technology that represents the opportunity cost to the national economy of providing that technology. There are three basic principles of economic costing:

- all relevant costs must be included;
- each cost must be properly evaluated; and
- the assumptions used for costing different technologies (especially those providing different levels of service) must be mutually consistent.

12.1.1 Inclusion of all relevant costs

The first principle, that *all* relevant costs must be included, is of fundamental importance. The key point is that all costs to the national economy, *regardless of who incurs them*, have to be included. An example of costs that have been conventionally ignored, and wrongly so, are household costs; for example, in determining the costs of conventional sewerage, engineers generally ignore not only the cost of the house connection, which is paid for by the householder, but they also fail to take into consideration the money spent by the householder inside the house—the cost of the toilet itself, plumbing costs, proportional construction costs (= house construction costs per m^2 × floor area of toilet room, m^2) and the cost of the water used to flush the toilet. In a *financial* analysis of costs to the sewerage authority, such an exclusion of household costs would be appropriate, but for an *economic* comparison of sanitation alternatives it is necessary to include *all* costs of each alternative regardless of whether they are borne by the household, the sewerage authority, the national government (which might give a grant or low-cost loan to the authority), or whomsoever. However, some financial costs should be ignored; for example, any taxes or subsidies, since these are only a transfer of money within the economy and not a cost to it.

12.1.2 Shadow pricing

The second principle of economic costing is the proper evaluation of each of the costs that have been included. How are these costs valued? The price actually paid for them (what economists call their market price) is not, in general, a good estimate of their opportunity cost to the national economy. Again, a good example of this is water: in a country with scarce water resources water will have a high economic cost, no matter how much or how little consumers actually pay for the water they consume. It is only by using prices that reflect actual resource scarcities that one can compare sanitation technologies in terms of their costs to the natural resource endowment of a country, i.e. their costs in terms of land, labour and capital.

National governments often have sociopolitical goals that may be only indirectly related to economic objectives, and one result of this is that some market prices may bear little relation to real economic costs. Because of this, it is generally necessary to adjust market prices when doing an economic costing exercise, so that they more closely represent 'real' unit costs in the sense of more closely reflecting their effect on the national economy. This adjustment of market prices to reflect opportunity costs is known as *shadow pricing*.

Market prices are converted to shadow prices (i.e. economic costs) by multiplying them by a conversion factor called the *shadow rate*. There are four shadow rates that are used in the economic costing of sanitation technologies:

- the unskilled labour wage shadow factor
- the foreign exchange shadow factor
- the opportunity cost of capital, and
- the shadow price of land, water and other direct inputs.

The calculation of the value of these shadow factors is very complicated and requires detailed knowledge of a country's economy. Those involved in sanitation programme planning are best advised to obtain the appropriate values from the Ministry of Economic Development, the World Bank or the African or Asian or InterAmerican Development Bank.

Unskilled labour

Many national governments enact minimum wage legislation in order to prevent exploitation of the labour force by unscrupulous employers. However desirable this may be in sociopolitical terms, it has the effect of economically *over*valuing unskilled labour: the pay an unskilled labourer receives is higher than he would receive if there were no minimum wage legislation. If a country has a very large pool of unemployed unskilled labourers, the shadow factor for unskilled labour wages would be close to zero as there is almost no cost *to the national economy* of employing such people, since they would be otherwise unemployed and so not contributing anything to the country's economic growth. On the other hand, if a country has a shortage of unskilled labour, the shadow factor would be 1, i.e. the market wage is a fair reflection of the economic cost of employment. Generally, in developing countries the shadow factor for unskilled labour is in the range 0.5–1.

Foreign exchange

Many national governments do not allow their national currency to be traded freely in the international money markets (to 'float'). Instead they fix its value, usually in terms of the currency of a major trading partner, for example the US dollar or the Japanese yen. This often results in the currency being *over*valued: imports cost less in terms of the national currency, and exports earn more foreign currency than if the national currency were traded freely on the international money markets. The foreign exchange shadow factor is the ratio of the shadow exchange rate (what the rate would be on the international money markets) to the official exchange rate fixed by the government (or the central bank); thus, its value is greater than (or equal to) 1. If, for example, the official rate of exchange is 15 units of national currency ('unc') per US$, but in the free market 25 unc would be required to purchase US$ 1, the foreign exchange shadow factor would be (25/15), i.e. 1.67.

Consider the cost of importing a high-performance vacuum tanker to be used to empty dry VIP latrine pits. Suppose the delivered cost of the tanker is US$ 125 000 (section 7.5). Using the official exchange rate of 15 unc per $, the municipality would have to pay 1 875 000 unc, but the economic or 'shadowed' cost to the country's economy is 1.67 times this amount, i.e. 3 125 000 unc, and this is the cost that should be used in evaluating the economic cost of the VIP latrine pit emptying service that the municipality wishes to implement.

At this point it is worth noting that (a) minimum wage legislation *over*values unskilled labour and (b) fixing the official exchange rate at too high a value *under*values imports. Thus, options that do not overly rely on labour but which require imported equipment are favoured. This is exactly the opposite of what developing countries need, as they have plenty of labour but little foreign exchange. Shadow pricing, which corrects the government sociopolitical distortions to the free market, restores the balance: labour-intensive schemes with low requirements for foreign exchange become more favoured.

Opportunity cost of capital

The opportunity cost of capital (OCC) is defined as the marginal productivity of additional investment in its best alternative use. The OCC can be thought of as the price (or yield) of money. In industrialized countries where capital is relatively abundant, its yield is relatively low (this is because capital has already been employed in its most productive uses and is now, in general, being substituted for labour (or other inputs) in less and less profitable areas). But in developing countries, capital is a scarce commodity and has a high opportunity cost. The government might decide, for good sociopolitical reasons, to make available loans to householders at a low rate of interest so that they can build, say, pour-flush toilets. The economic cost of this decision is the yield that the government would have received if it had invested its capital in the best alternative way—for example, by buying shares in a well-managed, highly-profitable industrial enterprise. The

OCC is thus expressed as a percentage, and in developing countries it is usually in the range 10–25 percent.

Land, water and other direct inputs

The costs of some inputs to a sanitation programme are often controlled by governments or incorporate government subsidies. For example, the land required for the construction of waste stabilization ponds (Chapter 10) may be owned by the government, and the government may decide to give it to the sewerage authority at no financial cost. However, its *economic* cost should be calculated as what it would have been worth had the government sold it to an industry (or farmer) that wished to locate there, and an idea of this value can be obtained from recent sales of land in the area.

Other costs that often need adjustment to reflect real resource costs are those of publicly produced outputs such as water and power. Their free market prices are rather difficult to estimate as the government normally has a monopoly in their production, and the shadow price of water or electricity has to be approximated by calculating the *average incremental cost* (*AIC*) of its production (see subsection 12.1.3.).

12.1.3 Mutually consistent assumptions

The third principle of economic costing requires that the assumptions used for costing different sanitation technologies are mutually consistent. Thus, one should not compare VIP latrines with, say, conventional sewerage because the latter accepts sullage but the former does not. The true comparison is between conventional sewerage and VIP latrines with sullage soakaways.

A further difficulty arises with the design population. For example, a VIP latrine has certain component costs that are independent of the number of users (the cost of the cover slab, superstructure, vent pipe and fly screen), but other costs that are dependent upon the number of users (pit volume, and hence, also, emptying costs). Thus, the per caput cost of a VIP latrine designed for a household size of x will not be the same as that of one designed for a household size of y. For this reason, sanitation costs should be expressed per household, rather than per caput.

Another difficulty with design population occurs because some sanitation facilities are fully used by their 'design population' as soon as they are constructed; for example, a VIP latrine will be used by all household members as soon as it is built, but others are not; for example, conventional sewerage is designed not just for today's population but for the population expected at the end of the design period, which is many years (perhaps 10 or 20) into the future. However, all the construction costs of conventional sewerage are incurred now, but its benefits (wastewater volumes collected) are realized only gradually over time. Just as costs incurred in the future have a lower present value than those incurred today, benefits received in the future are less valuable than those received today. Thus, serving a person (or a household or a sanitation basin) in n years' time is not worth as much as serving the same person (or household or sanitation basin) now. So to divide the investment cost of a conventional sewerage system by its design population would greatly *under*estimate its real per household cost when compared with that of a sanitation system that is fully used upon construction. (This underestimate becomes even greater when the in-house costs borne by the householders are ignored as well—subsection 12.1.1.)

The best way to overcome this difficulty with sanitation systems that are not fully utilized on completion is to calculate the average incremental cost (AIC) of the system per household. The AIC is simply the economic cost of the system per household per year. It is calculated as the sum of the present value of the shadowed construction costs and shadowed operation and maintenance costs, divided by the sum of the present value of the incremental number of households served. It is expressed in unc per household per year, and it is formally given by the equation:

$$AIC = \frac{\sum_{t=1}^{t=T} (C_t + O_t)/(1 + r)^{t-1}}{\sum_{t=1}^{t=T} N_t/(1 + r)^{t-1}} \tag{12.1}$$

where t = time, years
T = design life, years (measured from the start of the project at $t = 1$)
C_t = shadowed construction costs incurred in year t

O_t = shadowed operation and maintenance costs
N_t = cumulative number of households served in year t
r = opportunity cost of capital, percent/100

Equation (12.1) (which only looks complicated) is not, in practice, used, as one sets up a table of future costs and benefits that is then used to set up a second table of the present value of these future costs using the equation:

$$PV_t = FC_t/(1 + r)^{t-1} \quad (12.2)$$

where PV_t = present value of FC_t
FC_t = future cost incurred in year t

(A good way of understanding equation (12.2) is to realize that it tells you how much money (PV_t, unc) you must put into a bank account now, which, at an annual interest rate of $100r$ percent, will enable you to be able to pay out FC_t unc at the end of t years.)

Worked Example 6 in section 12.5 shows how equations (12.1) and (12.2) are used to calculate the *AIC* of conventional sewerage. Example 3 shows how the same approach can be used to calculate the economic cost of a VIP latrine that is mechanically emptied every three years; and Example 4 illustrates how the method can be used to estimate the costs of a mechanical pit-emptying service.

For a sanitation system that is fully utilized upon completion, the economic cost calculations are simpler, as all construction costs are incurred in year 1, and the operation and maintenance costs are the same each year. First, the construction costs are shadow-priced by using the appropriate shadow factors (for unskilled labour and foreign exchange) and then they are annuitized (over the design life of the system) using the opportunity cost of capital. Annuitization is simply a way of calculating an annual cost which, when multiplied by the design life in years, ensures that the capital sum and interest are both fully paid off by the end of the life of the system. The shadowed construction costs (C, unc) are annuitized by multiplying them by a capital recovery factor (*CRF*), which is given by:

$$CRF = [r(1 + r)^T]/[(1 + r)^T - 1] \qquad (12.3)$$

Thus, the total annual economic cost (*TAEC*, unc) of the system is given by the sum of the annuitized shadowed construction costs and the annual shadowed operation and maintenance costs (*O*, unc):

$$TAEC = (C \times CRF) + O \qquad (12.4)$$

Worked Example 1 in section 12.5 shows how equations (12.3) and (12.4) are used in practice to calculate the economic cost of a VIP latrine.

12.2 FINANCIAL COSTING

Economic costs are useful as they determine what the true costs to the natural resource endowment of the country are. But they are not especially useful to householders who are only interested in how much they have to pay for their VIP latrine, for example, or simplified sewerage, and how these costs will be spread over time. Nor are they useful to the sewerage authority, which has to determine how much to charge each household in order to maintain its own financial viability.

However, the problem with *financial* costs is that they are entirely dependent upon policy variables that can vary widely. Whereas economic costs reflect actual resource scarcities and are, therefore, quite objective, financial costs are entirely subjective in the sense that they are subject to loan policy (interest rate, length of loan), subsidies from the rich to the poor, and so on. For example, the government might decide (perhaps in order to increase its popularity) that water is to be free for everyone; so the financial cost of water to the consumer is zero (this is, in fact, a good example of the distinction between financial costs and economic costs: even if water has a financial cost of zero to the consumer, it still has an economic value to the country as a whole). Similarly, if a government decides to pay for sanitation out of central taxes, the financial cost to the householder is again zero. So financial costs by themselves cannot be used to make judgements between alternative sanitation technologies.

With on-site sanitation systems (VIP latrines or pour-flush toilets with sullage soakaways, or septic tanks), financial costs are determined in much the same way as economic costs, except that costs are not shadow-priced and the actual interest rate of any loan made available to the householder is used instead of the opportunity cost of capital (see Example 3 in section 12.5).

Financial costs of sewerage are, in practice, worked out differently, and what follows is equally applicable to conventional and unconventional (settled and simplified) sewerage. Water and sewerage authorities normally charge for sewerage by adding a sewerage surcharge to the water bill. Ideally, this surcharge should be based on the financial *AIC* of sewerage (calculated using costs that are not shadow-priced and the actual rate of interest on any loan the authority has obtained from, for example, the central government) and the sewage return factor (= m^3 of sewage produced per m^3 of water consumed; usually 0.80–0.95). Thus, the sewerage surcharge (unc per month or quarter) should be given by:

(water consumption, m^3 per month or quarter)

× (sewage return factor)

× (financial *AIC* of sewerage, unc per m^3)

In practice, water and sewerage authorities simply apply a straightforward percentage surcharge–for example, for conventional sewerage the surcharge is often 100 percent, i.e. consumers pay the same again for sewerage as they pay for water. However, given that conventional sewerage is usually two to three times more expensive than the water supply, the surcharge should be 200–300 percent; but anything over 100 percent is often politically difficult (the rich would complain vociferously) and this can lead to water and sewerage authorities having major problems of financial viability. It is not unknown for them to have to take out new loans to repay existing loans, or increase water charges (see section 1.3).

In the case of unconventional sewerage, the surcharge should be less than 100 percent: for example, in the city of Natal in northeast Brazil where simplified sewerage was developed (Chapter 9), capital costs in 1981 were US$ 325 per household and the water and sewerage company was able to recover its costs over a 30-year period by surcharging the water bill by only 40 percent. The charge for water was the 'minimum tariff' (i.e. an assumed unmetered consumption of 15 m^3 per household per month) of US$ 3.75, so the financial costs of simplified sewerage were really low, only US$ 1.50 per household per month.

12.3 HOUSEHOLD DECISIONS

Householders are interested in what they have to pay for their sanitation system, but what benefits can they expect to get? In economic terms, benefits are generally unquantifiable, but at the household level some reasonable assumptions can be made. For example, household members who are employed might spend less time off work due to a reduction in sickness due to excreta-related diseases, and they might have to spend less on medical drugs to cure excreta-related diseases. These benefits can be quantified reasonably well and, provided they are not overestimated, a fairly reasonable benefit–cost ratio can be established for the sanitation system (see Worked Example 3 in section 12.5).

12.4 FURTHER READING

G. A. Bridger and J. T. Wimpenny, *Planning Development Projects: A Practical Guide to the Choice and Appraisal of Public Sector Investments*. Her Majesty's Stationery Office, London (1983).

A. Cotton and R. Franceys, "Infrastructure for the urban poor in developing countries". *Proceedings of the Institution of Civil Engineers: Municipal Engineer* **98**, 129–138 (1993).

S. R. Curry and J. Weiss, *Project Analysis for Developing Countries*. Macmillan, London (1993).

D. W. Pearce and C. A. Nash, *The Social Appraisal of Projects*. Macmillan, London (1981).

A. Ray, *Cost-Benefit Analysis: Issues and Methodologies*. Johns Hopkins University Press, Baltimore, MD (1984).

R. Saerbeck, "Economic appraisal of projects: guidelines for a simplified cost benefit analysis". *EIB Papers No. 15*, pp. 59–78. European Investment Bank, Luxembourg (1990).

M. F. Scott, J. D. MacArthur and D. M. G. Newbery, *Project Appraisal in Practice*. Heinemann, London (1976).

I. Serageldin, *Water Supply, Sanitation and Environmental Sustainability: The Financing Challenge*. The World Bank, Washington, DC (1994).

L. Squire and H. G. van der Tak, *Economic Analysis of Projects*. Johns Hopkins University Press, Baltimore, MD (1975).

W. A. Ward and B. J. Deren, *The Economics of Project Appraisal: A Practitioner's Guide*. Economic Development Institute, The World Bank, Washington, DC (1991).
J. T. Wimpenny, *Values for the Environment: A Guide to Project Appraisal*. Her Majesty's Stationery Office, London (1991).

12.5 WORKED EXAMPLES

12.5.1 Example 1: Economic costing of a VIP latrine

1. Assume that:

   unskilled labour shadow factor = 0.6
   foreign exchange shadow factor = 1.9
   opportunity cost of capital = 18 percent

   latrine lifetime = 20 years

   skilled labour costs = 20 unc
   unskilled labour costs = 90 unc
   local materials costs = 120 unc
   imported materials costs = 50 unc

   operation and maintenance costs (including manual emptying costs) = 20 unc per year

2. Calculate the capital recovery factor from equation (12.3):

$$\begin{aligned} CRF &= [r(1 + r)^T]/[(1 + r)^T - 1] \\ &= [0.18(1.18)^{20}]/[(1.18)^{20} - 1] \\ &= 0.187 \end{aligned}$$

3. Set up Table 12.1. The shadow price (column 4) is the cost (column 2) multiplied by the shadow factor (column 3). The shadow prices are added together to give a subtotal (here, 289 unc), which is multiplied by the *CRF* (column 5) to give the annuitized shadowed construction costs (column 6; here, 54 unc). To this is added the annual operation and maintenance costs (20 unc), to give the total annual economic cost (*TAEC*) of the VIP latrine (column 6; here, 74 unc).

Table 12.1 Annual economic costs of a VIP latrine

(1) Item	(2) Cost (unc)	(3) Shadow factor	(4) Shadow price (unc)	(5) *CRF*	(6) Annual cost (unc)
Labour					
Skilled	20	–	20		
Unskilled	90	0.6	54		
Materials					
Local	120	–	120		
Imported	50	1.9	95		
Subtotal			289	0.187	54
Maintenance	20	–	20		20
Total					74

4. The *TAEC* can be expressed, if so desired, in US$ by converting at the *official* rate of exchange (because the costs have already been shadow-priced). Assume that the rate is 2.3 unc per $, so the *TAEC* of the VIP latrine is US$ 32.

Note: A rather better way of handling the emptying costs is shown in Examples 3 and 4.

12.5.2 Example 2: Financial costing of a VIP latrine

What would be the financial cost of the VIP latrine detailed in Example 1 if the householder provided the unskilled labour on a self-help basis, and was able to obtain a low-cost loan at 3 percent interest?

Table 12.2 sets out the costs. It is much simpler than Table 12.1 as the costs are not shadow-priced. The capital recovery factor for $r = 0.03$ and $T = 20$ years is, from equation (12.3), 0.067. So the annuitized construction costs are 12.70 unc; add to this the annual operation and maintenance cost of 20 unc, and the annual financial cost of the VIP latrine is 32.7 unc, i.e. US$ 14.22.

Table 12.2 Annual financial costs of a VIP latrine

(1) Item	(2) Cost (unc)	(3) *CRF*	(4) Annual cost (unc)
Labour[a]			
Skilled	20		
Unskilled	–[b]		
Materials[a]			
Local	120		
Imported	50		
Subtotal	190	0.067	12.70
Maintenance	20		20.00
Total			32.70

[a]Actually it is not necessary in a financial analysis to separate labour and materials costs as shown. It is done here for ease of comparison with Table 12.1.

[b]These are zero as the householder provides the unskilled labour on a self-help basis.

12.5.3 Example 3: Economic costing of a VIP latrine with mechanical emptying costs incurred every three years

Lumping the latrine manual emptying costs together with its annual maintenance costs, as done in Example 1, is really only valid if the latrine is emptied every 12 months. If it is mechanically emptied every three years, then it is more appropriate to set up a table of future costs as shown in Table 12.3 and explained, column by column, below.

Column (1): Year

This gives the year in which the costs listed in columns 2–4 are incurred, from year 1 (the year of construction; 1995, say) to year 20 (the end of the latrine's design lifetime, 2014).

Column (2): Shadowed construction costs

The figure of 289 unc shown against year 1 is the shadowed construction subtotal cost given in Table 12.1.

Table 12.3 Economic costs of a VIP latrine emptied every three years

(1) Year	(2) Construction	(3) Maintenance costs (unc)	(4) Emptying costs (unc)	(5) Total future costs (unc)	(6) *PVCF* (18%)	(7) Present value of future costs (unc)
1	289	5		294	1.000	294.0
2		5		5	0.847	4.2
3		5	69	74	0.718	53.1
4		5		5	0.609	3.0
5		5		5	0.516	2.6
6		5	69	74	0.437	32.3
7		5		5	0.370	1.9
8		5		5	0.314	1.6
9		5	69	74	0.266	19.7
10		5		5	0.225	1.1
11		5		5	0.191	1.0
12		5	69	74	0.162	12.0
13		5		5	0.137	0.7
14		5		5	0.116	0.6
15		5	69	74	0.099	7.3
16		5		5	0.084	0.4
17		5		5	0.071	0.4
18		5	69	74	0.060	4.4
19		5		5	0.051	0.3
20		5		5	0.043	0.2
				Totals:	6.316	440.8

Average incremental cost of VIP latrine = 440.8/6.316 = 69.8 unc per year

Column (3): Maintenance costs

An annual expenditure of 5 unc on cleaning materials is assumed for years 1–20 (i.e. construction is completed soon after the start of year 1.

Column (4): Emptying costs

These are only incurred every three years, towards the end of years 3, 6, 9, 12, 15 and 18. The shadowed cost of each emptying operation is taken as 69 unc (see Example 4).

Column (5): Total future costs

This column sums up for each year the costs given in columns (2)–(4).

Column (6): PVCF

This gives the 'present value conversion factor' (*PVCF*), which is simply the reciprocal of the denominator of the right-hand side of equation (12.2), i.e.

$$PVCF = (1 + r)^{1-t}$$

Here $r = 0.18$ and t = the year number in column (1).

Column (7): Present values

The future cost sums given in column 5 are now converted, year by year, to their corresponding present values by multiplying by the *PVCF* given in column 6. These present values are now summed to give, in this case, 440.8 unc.

Benefits

The benefit each year can be thought of as the latrine being in service for 1 year. So we could have a column (7) that would give the figure 1 for each of the 20 years. This would be multiplied by the *PVCF* in column 5 to give the present value of the year in service. But, of course, $1 \times PVCF = PVCF$, so the sum of the present values of the year in service is simply the sum of the *PVCF* values given in column 6 (here, 6.316).

All this is not as strange as it may first appear: having the latrine in service for 12 months in t years' time is worth less (i.e. has a lower present value) than having it in service for 12 months now.

Thus, the annual economic cost of the VIP latrine system is the sum of column (7) divided by the sum of column (6): here, 440.8/6.316, i.e. 69.8 unc or, converted at the official rate of exchange, US$ 30.3.

12.5.4 Example 4: Economic costs of mechanical pit emptying

Take the delivered cost of a BREVAC tanker as US$ 125 000, as noted in section 7.5. The shadow foreign exchange rate is 4.37 unc per $ (based on the official rate of exchange of 2.3 unc per $ and the foreign exchange shadow factor of 1.9, as used in Examples 1 and 3), so the shadow price of the tanker is 546 000 unc.

Assume that the tanker's annual operation and maintenance costs of 34 000 unc (= US$ 10 000 at the *official* rate of exchange; section 7.5) are broken down as follows:

Labour	
Skilled (i.e. the driver, vehicle mechanics)	5000
Unskilled	4000
Material	
Local	–
Imported (i.e. fuel, lubricants, spare parts)	14 000

These costs are shadow-priced, as in Example 1. The total shadowed annual operation and maintenance cost is thus 34 000 unc. A future costs table is set up (Table 12.4), but only

Table 12.4 Economic costing of VIP latrine emptying by vacuum tanker

(1) Year	(2) Capital costs (unc)	(3) O&M costs (unc)	(4) Total future costs (unc)	(5) *PVCF* (18%)	(6) Present value of future costs (unc)	(7) No. of pits emptied	(8) Present value of no. of pits emptied
1	546,000	34 000	580 000	1.000	580 000	2000	2000
2	–	34 000	34 000	0.847	28 798	2000	1694
3	–	34 000	34 000	0.718	24 412	2000	1436
4	–	34 000	34 000	0.609	20 706	2000	1218
5	–	34 000	34 000	0.516	17 544	2000	1032
6	–	34 000	34 000	0.437	14 858	2000	874
7	–	34 000	34 000	0.370	12 580	2000	740
8	–	34 000	34 000	0.314	10 676	2000	628
9	–	34 000	34 000	0.266	9044	2000	532
10	–	34 000	34 000	0.255	7650	2000	450
				$\sum PV$ = 726 268 (costs)		$\sum PV$ = 10 064 (benefits)	

for years one to ten, as the tanker costs are taken as written off after 10 years. Table 12.4 also includes the future benefits—the number of pits emptied each year (taken as 2000, as in section 7.5).

Next, the present values of these future costs and future benefits are determined, as in Example 3. Note that the present values of the future benefits are calculated in the same way as the present values of the future costs, using the same *PVCF*, i.e. using the opportunity cost of capital as the discount rate for the benefits (see Example 5).

Table 12.4 assumes that the tanker is delivered and starts emptying latrine pits at the beginning of year 1. It also assumes that the tanker maintenance costs are the same each year; this would not, in practice, be the case: repairs are more likely to occur in later years than at the start.

The present values of the future costs, and the present values of the future benefits, are now summed to give $\sum PV(\text{costs})$ and $\sum PV(\text{benefits})$. The average incremental cost of pit emptying is calculated as:

$$\begin{aligned} AIC &= \textstyle\sum PV(\text{costs})/\sum PV(\text{benefits}) \\ &= 726\,268/10\,604 \\ &= 68.5 \text{ unc (US\$ 29.8) per pit emptying operation} \end{aligned}$$

12.5.5 Example 5: Household decisions

Suppose a householder is considering having installed the VIP latrine that has the annual financial cost of 32.70 unc (i.e. US$ 9.87) given in Table 12.2. Is it, in practice, worth his while to spend this money? This question can only be answered by trying to estimate what benefits there might be. Suppose that he is now off work for two weeks less, so does not lose two weeks' income of, say, 14 unc per week; and he spends, say, 25 unc per year less on anti-diarrhoea drugs. So the latrine's benefits are (28 + 25) = 53 unc, and are thus greater than the cost of 32.7 unc. The benefit–cost ratio is (53/32.7), i.e. 1.6. Thus, even on this limited analysis it is certainly worth his while to go ahead and install a VIP latrine.

This is the sort of financial discussion that should be held with householders during the sanitation programme planning stage (Chapter 15).

12.5.6 Example 6: Conventional sewerage

Conventional sewerage has high capital costs during the construction phase, but does not normally provide its full benefits until the design population is reached later (usually much later) in the project's lifetime. As noted in subsection 12.1.3, the economic cost of such a system is best determined by calculating its average incremental cost (equation (12.1)). (However, an exception to this would be if conventional sewerage were installed in an urban area that was already at its saturation housing density, i.e. there is no space for any further houses to be constructed. This situation is often that encountered with settled sewerage and simplified sewerage. Economic costing, in this case, would be as in Example 1.)

In order to calculate the *AIC* of a conventional sewerage scheme, a future costs and benefits table is set up as shown in Table 12.5. The costs are considered under the following categories (each with its own column in Table 12.5):

- household capital costs
- household O&M costs
- collection capital costs
- collection O&M costs
- treatment capital costs, and
- treatment O&M costs.

All these costs have to be shadow-priced before they are entered into the table.

Household costs

The household capital costs are the cost of purchasing and installing a cistern-flush toilet, the cost of the house connection to the street sewer, and the construction cost of the toilet room (= house construction costs per m^2 × floor area of toilet room, m^2). Household O&M costs are principally the cost of the flushing water used per year; this is given by:

(AIC of water, unc/m^3)

$$\times \text{(number of times toilet flushed per day)}$$

$$\times \text{(flush volume, litres} \times 10^{-3} \times 365)$$

Collection and treatment costs

The capital costs of collection include all material and labour costs for the installation of the sewer network, including appurtenances such as manholes and pumping stations. Allowance should be made for replacing any pumps every few years. Collection O&M costs are often calculated as a small (0.5–1) percentage of capital costs.

Treatment costs are calculated similarly, not forgetting land costs. Treatment O&M costs are also usually calculated as a percentage (2–5) of capital costs.

Table 12.5 Future costs of conventional sewage scheme

Year	Household costs (unc)		Collection costs (unc)		Treatment costs (unc)		Sewage volume (10^3 m^3)
	Capital	O&M	Capital	O&M	Capital[a]	O&M	
1	280 000	0	1 762 800	0	5000	0	0
2	280 000	59 000	1 282 600	13 300	0	0	537
3	280 000	118 000	1 282 600	26 600	0	0	1073
4	280 000	177 000	1 282 600	39 800	0	0	1610
5	280 000	236 000	1 282 600	53 100	0	0	2146
6	0	295 000	0	66 400	0	0	2683
7	|	|	|	|	0	0	|
8	|	|	|	|	0	0	|
9	|	|	|	|	0	0	|
10	|	|	|	|	0	0	|
11	|	|	|	|	1 461 000	0	|
12	|	|	|	|	0	32 200	|
13	|	|	|	|	|	|	|
⋮	↓	↓	↓	↓	↓	↓	↓
44	0	295 000	0	66 400	0	32 200	2683
45	0	295 000	0	66 400	0	32 200	2683

[a]Electromechanical equipment replacement costs of 50 000 unc incurred in years 21, 31 and 41 (included in calculations but not shown above).

Example

The following example is based on the conventional sewerage scheme adopted in part of a city in West Africa. The cost data were as follows:

1. Shadow factors:

Unskilled labour	0.80
Foreign exchange	1.15
Opportunity cost of capital	12.0 percent
AIC of water	0.22 unc per m^3

2. Household size = 27
   Compounds per ha = 13
3. The 350 ha site would be sewered in 70 ha lots, one per year for five years; service would commence in each lot in the year following construction. The trunk sewers would be constructed in year 1, and the treatment plant would be constructed in year 11 (in the interim raw sewage would be discharged into a local river), but 10 ha land for treatment works (ponds) would be purchased at 500 unc per ha in year 1.
4. Wastewater flow = 60 litres per caput per day (lcd)
   Flushwater consumption = 30 lcd
   O&M costs:

Trunk sewers	0.5 percent of capital cost p.a.
Other sewers	1.0 percent p.a.
Pumping station and treatment plant	2.4 percent p.a.

5. Household maintenance costs (other than flushwater costs) ignored.
6. Pumping station electrical and mechanical equipment replaced every 10 years.

A table of future costs was then drawn up, as shown in Table 12.5. Then, as was done in Examples 3 and 4, these future costs were summed for each year, and the summed costs multiplied by the *PVCF* (calculated with $r = 0.12$) to give present values of the future costs. These were then summed to give the $\sum(PV)$costs, here 9 771 431 unc. Similar calculations were done for the benefits—the volume of sewage collected per year—to give the $\sum(PV)$benefits, here 17 901 029 m^3. The *AIC* of sewerage was:

$$9\,771\,431/17\,901\,029 = 0.55 \text{ unc per m}^3,$$

or, on a per compound basis:

$$60 \times 10^{-3} \times 27 \times 365 \times 0.55 = 163 \text{ unc per year}$$

13

Technology Selection and Upgrading

13.1 TECHNOLOGY SELECTION

The available urban sanitation technology options are described in Chapters 3–9. How does one choose between them? Which technology is most suitable for which urban community? The possibly unhelpful answer, but actually the only meaningful one, is that the most suitable sanitation option is the one that is:

- cheapest
- socioculturally acceptable, and
- technically and institutionally feasible

Even so, how does one choose between, say, VIP latrines and simplified sewerage, or pour-flush toilets and septic tanks, or settled sewerage? A useful starting point is to consider the community's water supply service level, as some sanitation technologies are not compatible with some service levels—conventional sewerage with standpipes, for example. The three levels of water supply service are:

- hand-carried supplies (standpipes, wells)
- on-plot supplies (sometimes called yard taps or patio connections—basically one tap per household or compound), and
- multiple-tap, in-house supplies

and their compatibilities with the various sanitation options are basically as follows:

Hand-carried supplies:	VIP latrines and pour-flush toilets (but people must be willing to cary home the pour-flush water)
On-plot supplies:	pour-flush toilets, simplified sewerage, settled sewerage, and
Multiple-tap, in-house supplies:	septic tanks, simplified sewerage, settled sewerage.

But, for a community that obtains its water from public standpipes, for example, how does one choose between VIP latrines and pour-flush toilets? VIP latrines are cheaper but, if people use water for anal cleansing, they will probably prefer to pay the little extra and have pour-flush toilets. Obviously, this should be discussed with them during the planning stage of the sanitation project (Chapter 15).

From these sorts of considerations and the water–sanitation compatibilities, one can develop the sanitation technology selection algorithm shown in Figure 13.1. An algorithm like this one is useful in that it makes you answer questions that you may not have thought of (or had forgotten) and it helps guide you to select what is usually the most appropriate sanitation option. However, it should not be used blindly in place of engineering judgement, and in this respect the word "usually" in the last sentence is important as there may be a special local circumstance that might lead you to select an alternative technology to the one given by the algorithm. Having said that, Figure 13.1 is a good starting point, but you should *always* appraise the technology it selects, not only against the local physical situation (soil type, depth of the groundwater table, and so on), but also with the community —in other words, check out its physicocultural applicability. Finally, is it economically appropriate and financially affordable?

13.2 TECHNOLOGY UPGRADING

Once a sanitation technology has been installed, that is not necessarily the end of the matter. Of course, there is operation and maintenance—HLOM (section 1.2) for both on-site and off-site sanitation systems, supplemented by municipal operation and maintenance for the latter, and there is also monitoring and evaluation (section 15). But there is usually also upgrading—principally upgrading on-site sanitation systems eventually to an off-site system, to follow improvements in the water supply service level. Another important upgrading (important, that it, to the users) would be the conversion of a pour-flush toilet to operate as a low-volume cistern-flush toilet (subsection 4.1.1).

13.2.1 Upgrading a VIP latrine

A single-pit or alternating twin-pit VIP latrine can be simply upgraded to a pour-flush toilet by installing either a squat-pan and trap unit in the cover slab (which has to be cut to size) or a pedestal-seat unit on the cover slab, having previously filled up the pit with rubble. The unit is then connected, with small diameter pipework, to an adjacent newly excavated leach pit.

13.2.2 Upgrading a pour-flush toilet

A single-pit pour-flush toilet can be upgraded to settled sewerage by emptying the pit, sealing the pit base by mortaring the vertical joints of the lining, and installing a sanitary-tee outlet that is connected to the settled sewer with 75 mm diameter pipe. Sullage is discharged into the pit.

An alternating twin-pit system can be upgraded by sealing both pits, as above, and permanently blocking off one of the exits in the flow diversion chamber, such that the toilet wastewater is discharged only into one pit. This pit is now connected to the second pit via a sanitary-tee and 75 mm diameter pipe. The second pit receives sullage, as well as the overflow from the first pit, and it is connected, via a sanitary-tee and 75 mm diameter pipe, to the settled sewer.

13.2.3 Planned sanitation sequences

People will upgrade their water supply before their sanitation system because of the greater convenience—going from a hand-carried water supply to an on-plot supply is obviously

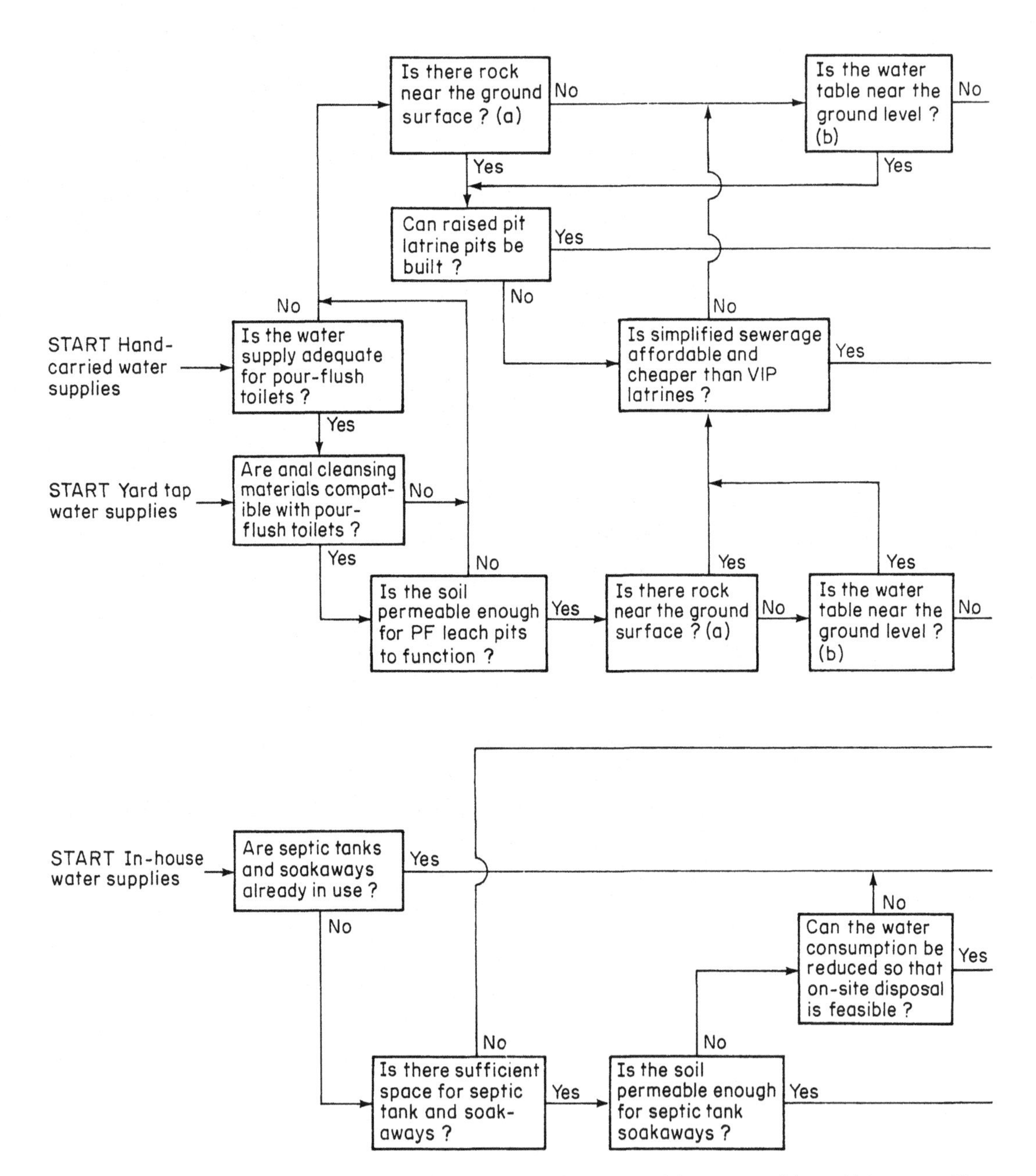

Figure 13.1 Sanitation technology selection algorithm. Notes: (a) to < 1 m; (b) to within 0.5 m either permanently or seasonally; (c) decide between single pits and alternating twin pits

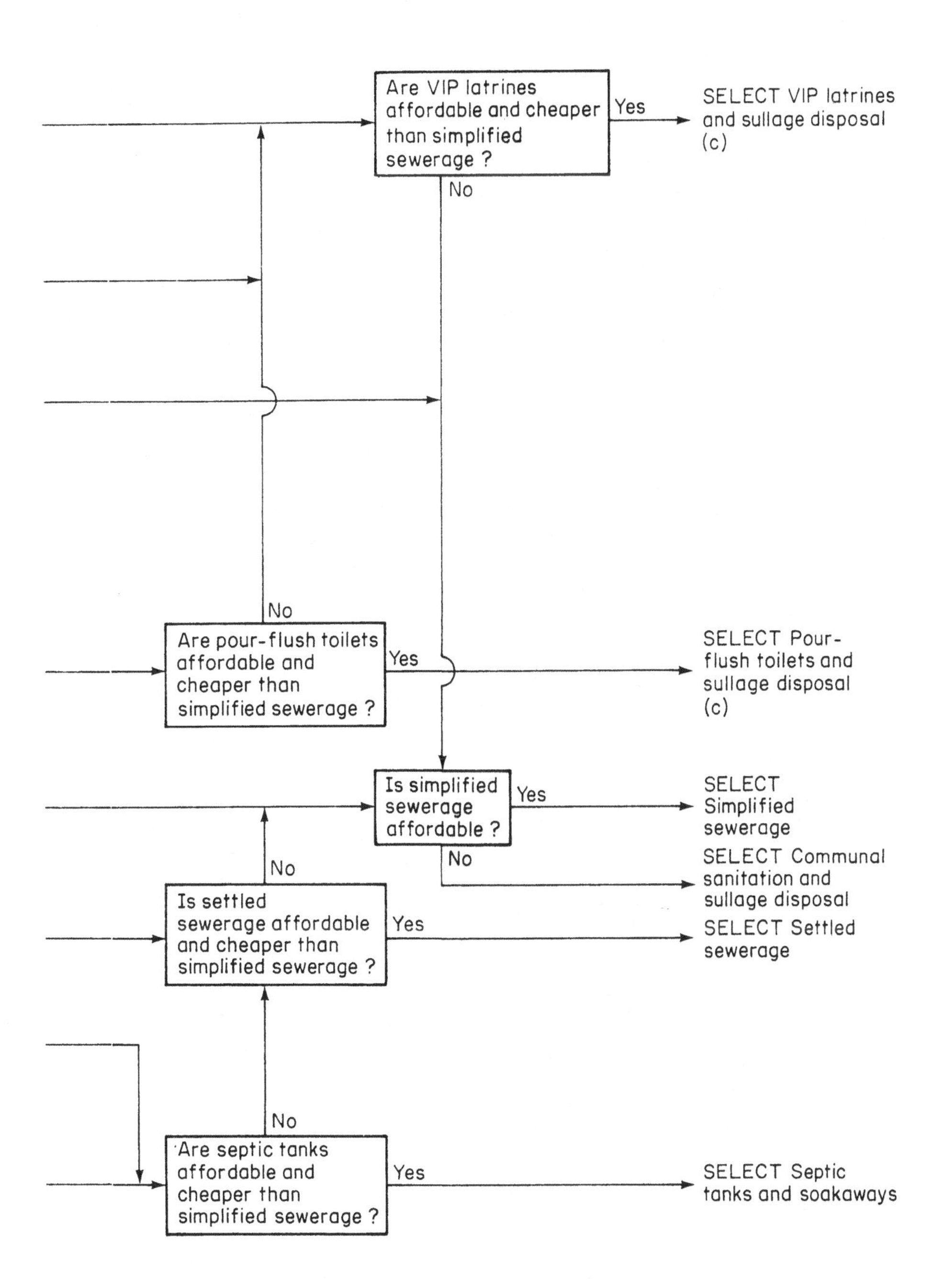
Are VIP latrines affordable and cheaper than simplified sewerage ?
Yes
SELECT VIP latrines and sullage disposal (c)
No
No
Are pour-flush toilets affordable and cheaper than simplified sewerage ?
Yes
SELECT Pour-flush toilets and sullage disposal (c)
Is simplified sewerage affordable ?
Yes
SELECT Simplified sewerage
No
SELECT Communal sanitation and sullage disposal
No
Is settled sewerage affordable and cheaper than simplified sewerage ?
Yes
SELECT Settled sewerage
No
Are septic tanks affordable and cheaper than simplified sewerage ?
Yes
SELECT Septic tanks and soakaways

much more convenient than changing one's VIP latrine to a pour-flush toilet or even upgrading the latter to a low-volume cistern-flush toilet. However, sanitation upgrading can be done as part of a planned sanitation sequence, as shown in Figure 13.2. Low-income urban communities can be served initially with, for example, VIP latrines or pour-flush toilets and hand-carried water supplies, and over time these can be upgraded to settled sewerage and multiple-tap, in-house supplies. This can be a very cost-effective way of providing urban sanitation: Table 13.1 shows that a 30-year planned sanitation sequence costs over eight times less than conventional sewerage, so over eight times as many people can be served via the former than with the latter: health for the many or convenience for the few? (In fact, Table 13.1 was based on relatively cheap—but still, of course, very expensive—conventional sewerage, and on relatively expensive—but still low cost—on-site systems; so probably over 10, rather than eight, times as many people could be served via the 30-year planned sanitation sequence. However, Table 13.1 was worked out before the development of simplified sewerage, which, as shown in Figure 9.2, can be cheaper than on-site sanitation.) An incidental advantage of Figure 13.2 and Table 13.1 is that

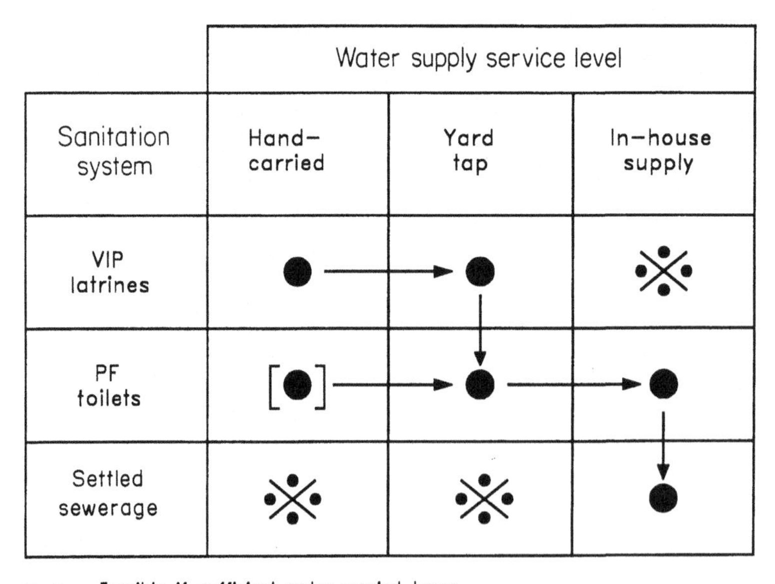

Feasible if sufficient water carried home for pour-flushing

Combination unlikely

Figure 13.2 Planned sanitation upgrading sequences

Table 13.1 Economic advantage of planned sanitation sequences

Sanitation system		Total present value cost per household[a] (1978 US$)	Relative cost (percent)
Planned sanitation sequence		354	12
Years 1–10	VIP latrine		
Years 11–20	PF toilet		
Years 21–30	Settled sewerage		
Settled sewerage (years 1–30)		1519	51
Conventional sewerage (years 1–30)		3000	100

[a]Capital and O&M costs discounted over 30 years at 8 percent (see section 12.1).

they can be used to persuade local politicians, especially those who say they only want conventional sewerage even for low-income communities, that there are better and more affordable ways of serving the urban poor.

14

Hygiene Education

14.1 INTRODUCTION

The mere provision of sanitation facilities, whilst essential for improved health, will not by itself achieve significant improvements in health. People may need improved water supplies as well. They almost certainly need to be shown how to use their improved water supply and sanitation facility to achieve maximal benefits to their health; this is called hygiene education (it used to be called health education, even sanitary education, but hygiene education is a more focused term). Hygiene education programmes seek to provide people with information that they can use to change their behavioural patterns so that they improve their health. Hygiene education is thus not coercive, but *motivational*. It is not simply telling people what to do, but rather explaining to them why they need to do it and how it will help them to help themselves to better health.

Hygiene is a somewhat general term that encompasses:

- *personal* hygiene—keeping ourselves clean: principally our bodies, especially our hands, and our clothes;
- *water* hygiene—keeping our water clean (protecting public standpipes from pollution (and vandalism), using clean containers to fetch, carry and drink water);
- *toilet* hygiene—keeping our sanitation facility clean (HLOM—section 1.2);
- *domestic* hygiene—keeping our home clean;
- *food* hygiene—keeping our food clean, so that it is safe to eat; and
- *peri-domestic* hygiene—keeping the areas around our house clean, and this includes safe garbage storage.

We should want to do all this to live in a clean environment, an environment that is substantially free from disease. As fully 80 percent of all morbidity in developing countries is due to water- and excreta-related diseases (section 1.1), having an adequate water supply and an adequate sanitation facility will be very important for our health, as will the knowledge we have to enable us to use these facilities in ways to improve our health to the greatest possible extent. Hygiene education seeks to provide this knowledge.

14.2 DISEASE EDUCATION

The 'germ theory' of disease may not be part of a community's common knowledge base. Diseases such as diarrhoea may be considered to be the manifestation of an angry god and, if people really believe this, then simply telling them that this is nonsense, that it is due to infection by a Group I or a Group II excreted pathogen, will not achieve much—except, perhaps, convincing them that *you* are crazy. People will have a clearer understanding of helminthic infections as dead adult worms are excreted from time to time, and this knowledge can be reinforced by chemotherapy to kill and expel *Ascaris* worms—see Figure 2.5. People who think that they are healthy often receive quite a shock when they see how many worms are expelled (in one campaign of this nature, they even ran a competition to see who expelled the greatest number of worms!). Sanitation promotion (sections 14.4 and 15.3) can consequently be much easier. The knowledge people have about helminths can be gradually extended to cover 'germs' —the Group I viruses and the Group II bacteria (section 2.2). The concept of faeco-oral transmission can be introduced, and this naturally leads on to the importance of water and sanitation in preventing such transmission—using a latrine or toilet to contain excreta and so prevent (or at least minimize) faecal contamination of one's immediate environment, and using water for keeping oneself and one's family, and one's environment, clean. We can think of this as maintaining a *water barrier* between ourselves and the Group I and II excreted pathogens.

14.3 HANDWASHING

Possibly the most important thing we do is to wash our hands *with soap* after using the toilet, after cleaning babies and toddlers when they have defecated, and before preparing and eating food. Numerous studies have shown that hand-washing is extremely important in interrupting the transmission of excreted pathogens, and people need to have this knowledge and to be able to put it into practice. So they need water and knowledge: but the water must be conveniently located—near the toilet or latrine, and near where food is prepared and eaten. Water storage facilities are therefore important, especially in households with a hand-carried water supply: can people, especially children, get sufficient water to wash their hands without contaminating the rest of the water?

Is soap available? If it is not, is ash available? Washing your hands with ash is as good as washing them with soap, and ash is commonly available from wood- or charcoal-burning stoves. (Washing your hands *without* soap or ash does virtually *nothing* to remove excreted pathogens, and so is an essentially useless exercise). Is soap affordable? Is there an opportunity here for local soap manufacture (see section 14.7)?

14.4 SANITATION EDUCATION

Sanitation education is education about the need for, and health advantages of, sanitation, and also education about how to use a particular sanitation system. The former can be linked with disease education (section 14.2), and is sometimes called sanitation promotion (see section 15.3). Education about a particular sanitation technology has to include all the HLOM requirements, as well as specific information—for example:

- *VIP latrines*: how they control odour and insects, why it is important not to cover the squat-hole;
- *pour-flush toilets*: the need to have adequate water for hand-flushing (and also for anal cleansing, if appropriate);
- *alternating twin-pit systems*: how they work, why only one pit should be used at a time;

- *on-site systems*: the need for emptying at regular intervals, and how this is to be done;
- *anal cleansing materials*: compatibility with the technology, advice on alternative materials.

14.5 COMMUNICATION

Interaction with a community can be easy and it can be difficult; it simply depends on the particular sociocultural setting. One point that is generally true, irrespective of the sociocultural setting, is that engineers are, by their very training, *not* the best people to interact with a community (there are of course, exceptions to this—but not many!). Professionals in sociology, anthropology and behavioural science are more appropriate, as are the local cadre of health workers, such as health inspectors and health assistants. Communication specialists, such as those working in advertising, also have a role to play in hygiene education and sanitation promotion (see section 15.3).

What media are best locally suited to communicate hygiene and sanitation education to an urban audience? Mass media options include newspapers, magazines (especially women's magazines and children's comics), radio, television, posters, leaflets, even postage stamps (but stamps for local letters, rather than airmail stamps). Articles, phone-in programmes, quizzes, dramas are all possibilities, as are face-to-face activities such as local meetings, talks at the health clinic, and so on. But whatever is done, it must be comprehensible to the local people, not all of whom may be literate, and it must respect their cultural sensitivities. Pre-testing the message, so that it does reach its intended audience in a effective manner, is extremely important. And the message, once pre-tested and refined, needs to be repeated almost *ad nauseam*, so that (as advertisers will tell you) it becomes embedded in the target audience's subconscious.

It is sensible to *prioritise* hygiene education messages, so that one is *not* trying to do everything at once—this is simply too confusing. Probably the first two things to do are:

- promote handwashing with soap or ash (see section 14.3), as a means of controlling diarrhoeal disease, and

- demonstrate worm infections by worm expulsion from selected young children (see section 14.2 and Figure 2.5).

Who comprises the target audience? Women are the most important group: they, as mothers and daughters, are responsible for HLOM of the toilet or latrine, for domestic hygiene, food preparation, toilet-training young children and looking after babies and toddlers too young to use the toilet or latrine. So the programme of hygiene education must be specifically designed for the local women (but men's needs should not be ignored). This means that those who interact with the local women must be women (and preferably local women) themselves.

Almost anyone can build a VIP latrine, even thousands of them; but to make sure that they are actually used in a way that really does improve the health of all their users, is something else—something that requires a really effective and sustained programme of hygiene education. Such a programme has its own logistical requirements and funding needs: these must be adequately assessed, and budgeted for and provided (otherwise the programme will not achieve its aims).

14.6 EDUCATING CHILDREN

Children are most at risk from excreta-related disease, and they need to receive hygiene education messages specifically designed for them. Of course, they will receive some information from their mothers and elder sisters, and also at school (so the school toilets should be kept scrupulously clean—but often they are not). Preschool-age children may go to nursery school or crèches, and these can provide a good forum for hygiene education though appropriately designed play activities—puzzles, stories, jokes, songs, drawing, acting games and so on. School attendants may well need some simple training to enable them to organize such activities to maximize the hygiene education effect.

At school there is much obvious opportunity for health-related education: not just hygiene education and environmental sanitation but also, as part of biology or housecraft, for example, how some of the excreta-related diseases are transmitted, and the life cycles of some of the pathogens (*Ascaris*, for example and *Wuchereria bancrofti*). Eight-year olds are more than capable of understanding how a VIP latrine works, and why fly and odour control is important, and they enjoy making a cardboard cutout model of it. And they can tell their parents all about it when they get back home.

14.7 FURTHER READING

A. Almedon (ed.), "Behaviour change for the better". *Waterlines* **13**(3)—a whole issue on hygiene education (January 1995).

K. Attawall and K. Miles, "Making soap". Technical Brief No. 8, In: J. Pickford (ed.), *The Worth of Water*, pp. 32–35. IT Publications, London (1991).

I. C. Black, *Low-cost Urban Sanitation in Lesotho*. Water and Sanitation Discussion Paper No. 10. The World Bank, Washington, DC (1994).

M. Boot, *Making the Links: Guidelines for Hygiene Education in Community Water Supply and Sanitation*. Occasional Paper No. 5. IRC International Water and Sanitation Centre, The Hague (1990).

M. T. Boot, *Just Stir Gently: The Way to Mix Hygiene Education with Water Supply and Sanitation*. Technical Paper No. 29. IRC International Water and Sanitation Centre, The Hague (1991).

M. Boot and S. Cairncross (eds), *Actions Speak: The Study of Hygiene Behaviour in Water and Sanitation Projects*. IRC International Water and Sanitation Centre, The Hague (1993).

L. Burgers, M. Boot and C. van Wijk-Sijbesma, *Hygiene Education in Water Supply and Sanitation Programmes*. Technical Paper No. 27. IRC International Water and Sanitation Centre, The Hague (1988).

P. Donkor, *Small-scale Soapmaking: A Handbook*, IT Publications, London (1986).

A. M. Han and T. Hlaing, "Prevention of diarrhoea and dysentry by hand washing". *Transactions of the Royal Society for Tropical Medicine and Hygiene* **83**, 128–131 (1989).

J. Hubley, *Communicating Health: An Action Guide to Health Education and Health Promotion*. Macmillan, London (1993).

D. Nyamwaga and P.Akuma, *A Guide to Health Education in Water and Sanitation Programmes*. African Medical and Research Foundation, Nairobi (1986).

H. Perrett, *Planning of Communications Support (Information, Motivation and Education) in Sanitation Projects and Programs*. TAG Technical Note No. 2. The World Bank, Washington, DC (1983).

J. Waterkeyn, "Community mobilization: where is the entry point?" *Waterlines* **9**(4), 2–4 (1991).

15

Institutional Aspects

15.1 ORGANIZATIONS AND RESPONSIBILITIES

Whether the low-cost urban sanitation programme is concerned with new housing developments, including site-and-service schemes, or with upgrading existing housing estates, shanty towns or slums, or both, there are several key elements that it should include. These are listed in Table 15.1. On-site

Table 15.1 Key elements of a sanitation programme

Item no.	Element
1.	A central steering committee comprising the ministries or departments responsible for finance and planning, health, urban development, water supply and sewerage
2.	Sound project management, site investigations, careful technology choice and design
3.	Pre-programme study of social factors, economic constraints, and beneficiary preference
4.	Development of an extension system, including health education, technical assistance to self-building householders, and feedback from the community
5.	Access to and delivery of building materials and mass-produced components, combined with financing mechanisms
6.	Integration of designs with infrastructure development, particularly water supply, storm water drainage and housing layouts
7.	Integration of programme management with existing administrative structures, such as city or town councils
8.	A monitoring and evaluation programme
9.	A programme for briefing central government personnel, and for training engineers, technicians, artisans, and extension workers

Source: S. Cairncross and R. Feachem, *Environmental Health Engineering in the Tropics*, 2nd edition (Wiley, 1993).

sanitation programmes require rather more complex institutional arrangements than off-site sanitation programmes, and the following question has to be asked early on: is on-site sanitation actually cheaper than off-site sanitation (i.e. settled or simplified sewerage, including sewage treatment), for which the local water and sewerage authority is responsible? Figure 9.2 indicates that simplified sewerage is cheaper than on-site systems at high housing densities, so this must be carefully examined very early in the planning process.

Sanitation *programmes* are often national or provincial, and sanitation *projects* local. Ideally, sanitation projects form part of a sanitation programme, but this can be a source of institutional conflict, especially in the case of on-site sanitation. There is often mistrust between the national (or provincial or state) government and local governments: funds for projects generally come from the former, but the projects are usually planned and implemented by the latter. Local agencies are better placed to interact with the communities that will benefit from the projects, but their staff may need training so that project planning and implementation proceeds smoothly. Co-ordination and co-operation is vital.

15.2 SANITATION PLANNING

15.2.1 The Kalbermatten model

The old way of sewerage planning—long and expensive master plan studies, which resulted in a multi-volume detailed engineering report but few, if any, services for the urban poor—is no longer appropriate. The new model for sanitation planning, developed by John Kalbermatten and his colleagues at the World Bank in the late 1970s in preparation for the International Drinking Water Supply and Sanitation Decade, is shown in Figure 15.1. It involves several professional disciplines (not just engineering) *and the community* (see section 15.4). The professional disciplines are:

- public health engineering
- public health medicine
- economics

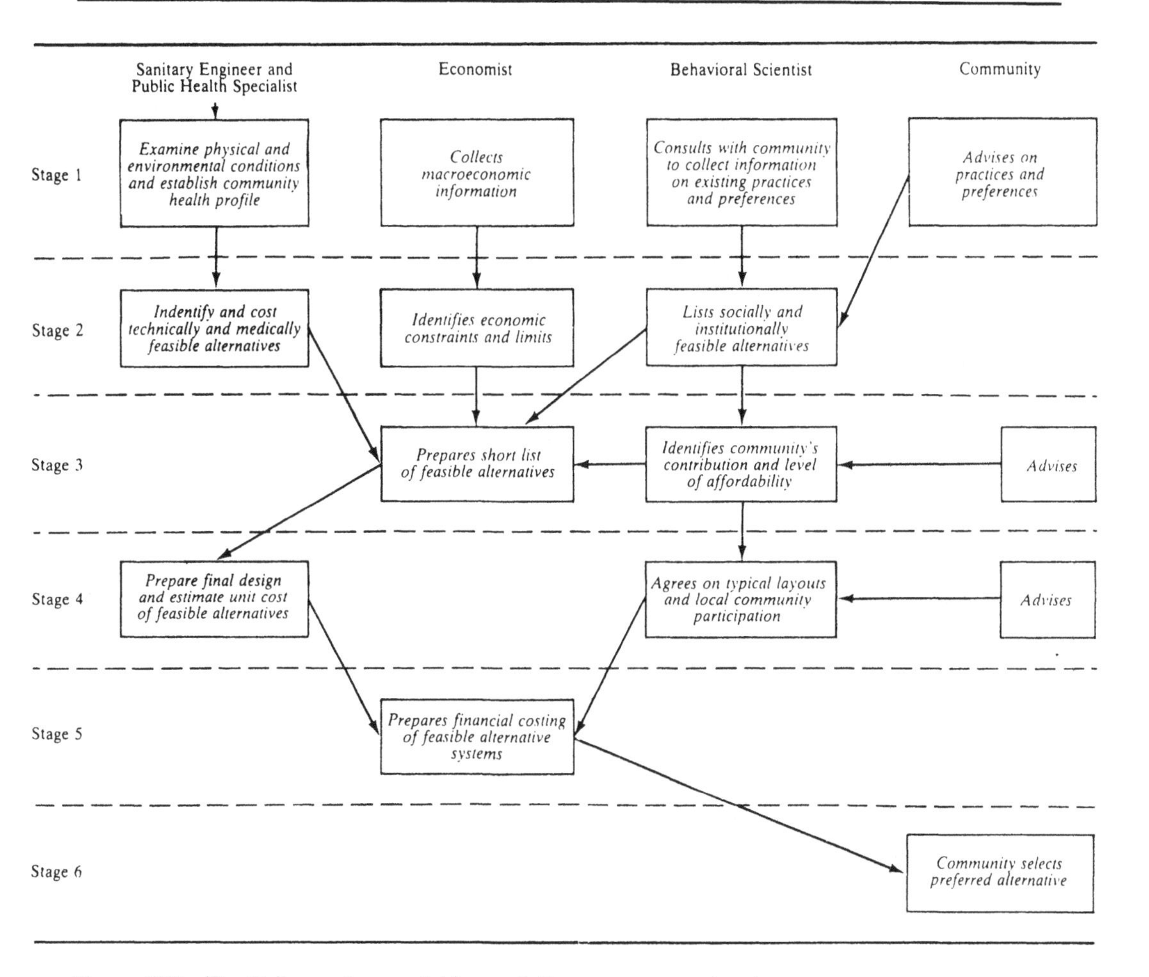

Figure 15.1 The Kalbermatten model for sanitation programme planning

- financial analysis, and
- behavioural sciences (such as sociology, anthropology).

The basic idea is to have an interdisciplinary programme or project team that can properly and sensitively interact with the community to produce technically feasible, economically and financially affordable, and socioculturally acceptable sanitation systems that will improve the community's health, reduce peri-urban pollution and raise the general quality of urban life in low-income areas.

The Kalbermatten model is just that—a model, and it should be adapted to suit local circumstances, rather than being blindly adopted. Successful low-cost urban sanitation programmes all contain the basic elements of the Kalbermatten planning model. Adopt it, adapt it, but ignore it at your peril!

In Stage I of the Kalbermatten model, engineers collect physical and environmental data (water supply details, including consumption; local topography, drainage, soil characteristics, position of the groundwater table, and so on); the public health specialist establishes which excreta-related diseases are prevalent in the community; the economist collects the macro-economic information needed for socioeconomic analysis; and the behavioural scientist and the community together determine the community's practices and preferences. This all enables the team to identify, in Stage II, the sanitation technologies that are technically, economically and socioculturally feasible. This then permits in Stages III–V, again with the help of the community, one or two of these technologies to be properly costed out in financial terms so that they are affordable. In Stage VI the community chooses the technology to be implemented: it is *their* choice, *their* technology.

Experience shows that a low-income urban community that has been involved in the planning of its sanitation facilities, is much more receptive and responsive to hygiene education. However, hygiene education should only start once the sanitation facilities are installed, *not* before, as the messages of hygiene education cannot be put into practice by the householders until they have their sanitation facilities.

15.2.2 Operation and maintenance planning

Planning for operation and maintenance (O&M), including training householders in HLOM, is vital to the success of low-cost urban sanitation projects. Institutional responsibilities must be clearly defined, and the necessary equipment obtained and associated training undertaken, for the following:

- emptying VIP latrine and pour-flush toilet pits
- desludging septic tanks and solids interceptor tanks (settled sewerage)

- sludge disposal
- maintenance of settled and simplified sewerage networks, including sewer flushing and unblocking, and lift stations, and
- sewage treatment and effluent reuse.

15.3 SANITATION PROMOTION

Sanitation promotion seeks to persuade householders to decide to install (or have installed) a sanitation facility, to pay for it, and to use, operate and maintain it in a way that brings health and environmental benefits. It thus comprises elements of 'social marketing' (information, motivation and education), and also of hygiene education in particular (Chapter 14). Social marketing uses some of the techniques of commercial marketing (advertising, for example), but it is aimed at the whole community, rather than individuals, so that the initial take-up rate is high (this is especially important in simplified sewerage schemes in which one has to aim for an initial connection rate of over 90 percent to ensure proper hydraulic functioning—see section 9.2). It is important to pre-test any advertising messages, motivational leaflets (can everyone read?), and so on, to make sure they are effective: feedback from one's 'customers', especially women, is very important.

15.3.1 Demonstration facilities

Community members may need to see the various sanitation options before they can decide which one they wish to have (section 15.2). Demonstration facilities are useful for this purpose: they can be inspected, even used, at will. However, if demonstration units are built at a place where, sooner or later, almost everyone goes—at the local health post, a school or crèche, church or mosque, or the local office of the national political office, for example—then the units have to be designed for such heavy use, and they will be slightly different (more robust, larger pits) from household units; but they should not be too different. With settled and simplified sewerage, of course, a demonstration unit is not so straightforward as it covers at least a whole housing block. Perhaps, in the office of the sanitation team, an extra toilet can be made available for inspection and the connecting sewer left exposed in its trench, so that people can see how the system is designed to operate.

15.3.2 Selecting communities

In urban areas, community selection can be problematic, especially if the sanitation promotion programme is effective; many low-income communities may then all be demanding improved sanitation facilities at the same time. Who chooses between them, and how should such a choice be made? This can become even more difficult if two competing agencies are involved—for example, the municipal sanitation department, which might be responsible only for on-site sanitation, and the local water and sewerage authority, which is concerned only with sewerage (but hopefully not just conventional sewerage, but also settled and simplified sewerage). As pointed out in sections 15.1 and 15.2, organizational responsibilities must be clearly delineated, inter-organizational competition avoided and the municipality-wide sanitation programme properly conceived and operated, such that community-level sanitation projects can be properly planned and implemented.

If the very poorest communities are selected first, then slightly less poor communities might reject the sanitation system selected for them if it is the same as that selected for the poorest communities, simply for that reason—it might be good enough for the very poor, but *not* good enough for *them*. So serving the very poor first is not, in practice, the best approach. It is better to serve the less poor initially, then the very poor (who may need subsidies—see section 15.5; and this is another reason for not serving them first, as the less poor would most probably demand subsidies as well).

15.4 COMMUNITY PARTICIPATION AND MANAGEMENT

Community participation is not simply the contribution of self-help labour (to excavate pits for VIP latrines or pour-flush toilets, for example, or simplified sewer trenches). This is, of course, important to reduce costs, but it has to be managed (supervised) to ensure its effectiveness.

As noted in section 15.2, the community must be involved from the beginning; their views on a range of topics must be sought and incorporated into the sanitation programme or project plan (what do they want? what can they afford? what are their sociocultural requirements?). An equal partnership must be established between the community and the appropriate governmental agency (the sanitation department of the municipality, for example, or the sewerage authority). The aim is for the latter to help the former to have the sanitation facilities it needs, and also to operate and maintain them properly.

A fully participating community makes the work of sanitation professionals much more interesting and effective: there is almost constant feedback, which is especially valuable not only at the planning and implementation stages, but also during monitoring and evaluation (section 15.6). The views of women are especially important and *must* be heard; in some cultures it may be necessary to have separate meetings for women (and the local agency staff attending these should be women as well). Effective community participation facilitates virtually everything—sanitation promotion, planning, implementation, hygiene education, operation and maintenance, and monitoring and evaluation; it maximizes the opportunities to achieve health benefits, to reduce peri-urban pollution and to give low-income communities what might be termed 'urban dignity'.

15.5 SUBSIDIES OR LOANS?

It has to be recognized that, for many low-income—and especially very low-income—communities, low-cost sanitation is not necessarily cheap. It is, of course, much cheaper than conventional sewerage, and it may be a good investment both economically and financially (see Chapter 12), but if people do not have the money to build (even self-build) their VIP latrine, for example, what are they to do? More to the point, what are *we*—as sanitation programme planners or sanitation project engineers—to do?

We have, perhaps, a choice: either subsidise the cost of the sanitation facility, or arrange for loans (even at, perhaps, subsidised rates of interest) to be made available.

15.5.1 Subsidies

The obvious point has to be made: subsidies cost money! Do we have the money, or can we get it? However, even if we have it, or can get it from some body that has it (central government or, more likely, a bilateral aid agency), should we use the money to make direct subsidies to households for construction of their sanitation facility? Maybe a better use of any such 'free' money would be to cover the overhead costs of a really effective hygiene education programme. Perhaps any direct subsidy should be limited to providing reduced rates of interest or loans (subsection 15.5.3) and/or some key component—the fly screen for a VIP latrine, for example, or the glass-fibre-reinforced plastic squat pan and trap unit for a pour-flush toilet, and selling these items to participating householders at half-price or less. What subsidies must *not* do is to remove or lessen the householders' feeling of ownership and responsibility: *their* sanitation facility belongs to *them*, and *they* must be responsible for its HLOM.

There is a case, however, for providing larger subsidies to extremely poor householders in the project area—single mothers with several young children, for example, or elderly abandoned widows. It is probably best to discuss the criteria for eligibility for these larger subsidies with the community as a whole, or its leaders, so that they are properly understood and do not lead to unjustifiable demands from less poor householders.

A further point is that any subsidies must not lead to tenants being penalized by their landlords: any rent increase should reflect *only* the costs incurred by the landlord in installing the sanitation facility. Often some form of control by the project office or local government (perhaps even by national legislation) is necessary to ensure that tenants are treated fairly.

15.5.2 Loans

Loans, possibly at a subsidised rate of interest, may need to be made available to householders to enable them to install their sanitation facility. Care should be exercised in setting the interest rate and the loan repayment term: are the monthly repayments affordable? Should the sanitation loan be made part of any existing housing loan?

Of course, some control is needed to ensure that the money lent is actually spent on sanitation. The procedure adopted in Lesotho by the Urban Sanitation Improvement Team (USIT, within the Ministry of Interior) for its very successful urban VIP latrine programme may serve as a model for other urban sanitation programmes; it is detailed in Table 15.2. Loans are provided not by USIT or the government, but by a commercial bank, the Lesotho Bank. Few poor families have failed to make their loan repayments (some wealthier families have, as they thought no-one would check up on them!), and this is attributed to several factors:

- The loan application is submitted to a Loan Approval Committee of two USIT officers and three or four local people such as the town clerk, chief, hospital matron or military officer. The LAC must interview the client before the loan can be approved. The LAC is also responsible, along with USIT, for following up on late repayments.

- A substantial deposit (30–40 percent of total cost) is required. This involves not only money but organization, time and effort and helps to ensure the client is serious about wanting the latrine.

- Lesotho Bank is regarded as an efficient institution that would certainly try to reclaim debts if they were owed.

- Reminders are given or sent out automatically after 30, 60 and 90 days when a repayment has been missed. The 90-day reminder is copied and circulated to the local chief (or town clerk) who will also follow up the debt.

- USIT strongly encourages people to visit the office to discuss repayment problems. Community staff follow up on clients who are defaulting on their loans.

- No repayments are expected during December and January when everyone has the costs of Christmas as well as school fees to pay. The loan is actually repaid in 20 instalments over 24 months.

Table 15.2 Procedure to be followed by householders in urban Lesotho who wish to obtain a loan for sanitation improvements.

Step no.	Activity
	How to get credit for your VIP
	If you do not have funds to build a VIP now, then you can apply to USIT for a loan from Lesotho Bank. This is what you have to do:
1.	Go to your nearest USIT office and ask for a full explanation of the Loan Scheme and the various options available.
2.	With USIT assistance, complete the Loan Application Form. You can choose to repay your loan over any period of up to 20 months. Interest will be charged at the normal Lesotho Bank rates on the loan. In exceptional circumstances, repayment of the loan could be negotiated for a longer period.
3.	You will then be called for an interview by the Loan Approval Committee (LAC). They need to check that you are over 18, that you can produce a site ownership certificate and that you are likely to meet your monthly repayments. Before you can receive the loan, you will have to collect 120 blocks and sand for the substructure. You must also dig your own pit.
4.	When you have collected the materials, you will have to sign an 'Acknowledgement of Debt' agreement and commit your collateral against the loan amount. You should then pay the M10.00 registration fee. This fee covers the cost of paperwork, fly screen, roofscrews and a few small items. You will then be given a 'loan number'.
5.	USIT will help you find a trained builder and give you a purchase order for the remaining materials and the builder's fee.
6.	After you collect the materials yourself from the suppliers, the builder can start building. A USIT Technical Officer will check that it is built correctly. When it is finished, you will have to sign a completion certificate, stating that you are satisfied with the VIP—before the builder is paid.
7.	When the invoices have been paid, USIT will set up the loan with Lesotho Bank. You will be given a Loan Repayment Card to take with you to the Bank. The repayment should be made on or before the first day of each month.
8.	If you have any financial problems and cannot make a repayment, talk to USIT community staff about it and USIT will try to help you. Remember, if you repay in less than 20 months, you will pay less money in interest.

Source: Blackett (section 15.7).
Notes: USIT, Urban Sanitation Improvement Team. M10.00 = US$ 3.70.

15.6 MONITORING AND EVALUATION

After a sanitation project has been implemented and operational for around a year, it should be monitored to see how it is working—is the design satisfactory? is everybody (all household members, *including* children) actually using the facility? has hygiene improved? has excreta-related disease decreased? and so on. The information so gathered is useful not only to correct any problems, but also to help design new projects better.

Sanitation programmes need evaluation, which is similar to monitoring but more comprehensive. Evaluation is concerned with:

- *functioning*: are improvements to physical design needed? If some facilities are non-functioning, why is this and how can they be made to function?
- *utilization*: what is the proportion of households covered (*coverage* = the number of households with a functioning sanitation facility, divided by the number of households in the programme (or project) area)? what is the *facility usage* (= number of household members who use the facility, divided by the total number of household members)?
- *impact*: what benefits have been achieved? improved health? improved hygiene? reduced peri-domestic, peri-urban, environmental pollution? how can these benefits be maximized?

The World Health Organization has provided a very useful minimum evaluation procedure for water supply and sanitation projects (see section 15.7), which concentrates primarily on functioning and utilization. Impact assessment is more complicated as the epidemiology is difficult, but an attempt should be made to determine the health and other benefits of a sanitation programme. Feedback from the community is essential.

15.7 FURTHER READING

G. J. Alaerts, T. L. Blair and F. J. A. Hartvelt (eds), *A Strategy for Water Sector Capacity Building*. IHE Report Series No. 24. International Institute for Hydraulic and Environmental Engineering, Delft (1991).

Better Urban Services: Finding the Right Incentives. The World Bank, Washington, DC (1995).

I. Blackett, *Low Cost Urban Sanitation in Lesotho*. Water and Sanitation Discussion Paper No. 10. The World Bank, Washington, DC (1994).

A. M. Cairncross, "Health impacts in developing countries: new evidence and new prospects". *Journal of the Institution of Water and Environmental Management* **4**(6), 571–577 (1990).

S. Cairncross, *Sanitation and Water Supply: Practical Lessons from the Decade*. Water and Sanitation Discussion Paper No. 9. The World Bank, Washington, DC (1992).

A. P. Cotton and W. K. Tayler, "Community management of urban infrastructure in developing countries". *Proceedings of the Institution of Civil Engineers—Municipal Engineer* **103**(4), 215–224.

J. W. Cusworth and T. R. Franks (eds), *Managing Projects in Developing Countries*. Longman Scientific and Technical, Harlow (1993).

Financial Management for Water Supply and Sanitation: A Handbook. World Health Organization, Geneva (1994).

W. F. Fox, *Strategic Options for Urban Infrastructure Management*. UNDP/UNCHS/World Bank Urban Management Program Paper No. 17. United Nations Centre for Human Settlements, Nairobi (1994).

C. Kessides, *Institutional Options for the Provision of Infrastructure*. Discussion Paper No. 212. The World Bank, Washington, DC (1993).

A. LaFond, *A Review of Sanitation Program Evaluations in Developing Countries*. EHP Activity Report No. 5. Environmental Health Project, Arlington, VA (1995).

R. Manoff, *Social Marketing: New Imperative for Public Health*. Praeger, New York (1985).

D. Marsden and P. Oakley (eds), *Evaluating Social Development Projects*. Development Guidelines No. 5. OXFAM, Oxford (1990).

Minimum Evaluation Procedure (MEP) for Water Supply and Sanitation Projects, Report No. ETS/83.1. World Health Organization, Geneva (1983).

P. Nichols, *Social Survey Methods: A Field Guide for Development Workers*. Development Guidelines No. 6. OXFAM, Oxford (1991).

P. L. Njoh, "Building and urban land use controls in developing countries". *Third World Planning Review* **17**(3), 337–356 (1995).

Operation and Maintenance of Urban Water Supply and Sanitation Systems: A Guide for Managers. World Health Organization, Geneva (1994).

I. Serageldin, M. A. Cohen and J. Leitmann (eds), *Enabling Sustainable Community Development*. Environmentally Sustainable Development Proceedings Series No. 8. The World Bank, Washington, DC, (1995).

L. Shrinivasan, *Tools for the Community Participation: A Manual for Training Trainers in Participatory Techniques*. PROWWESS/United Nations Development Programme, New York (1990).

The Maintenance of Infrastructure and Its Financing and Cost Recovery. United Nations Centre for Human Settlements, Nairobi (1989).

UNICEF/IRC, *Evaluating Water Supply and Sanitation Projects* (2 vols). Training Series No. 2. IRC International Water and Sanitation Centre, The Hague (1987).

R. C. G. Varley, *Financial Services and Environmental Health: Household Credit for Water and Sanitation*. Applied Study No. 2. Environmental Health Project, Arlington, VA (1995).

G. Watson, *Good Sewers Cheap? Agency—Customer Interactions in Low-cost Urban Sanitation in Brazil*. Water & Sanitation Currents. The World Bank, Washington, DC (1995).

D. Whittington *et al.*, *Household Demand for Improved Sanitation Services: A Case Study of Kumasi, Ghana*. Water and Sanitation Report No. 3. The World Bank, Washington, DC (1992).

World Bank, *World Development Report 1994: Infrastructure for Development*. Oxford University Press, New York (1994).

Index

9 780471 961635